Life Writing in the Anthropocene

Life Writing in the Anthropocene is a collection of timely and original approaches to the question of what constitutes a life, how that life is narrated, and what lives matter in autobiography studies in the Anthropocene. This era is characterised by the geoengineering impact of humans, which is shaping the planet's biophysical systems through the combustion of fossil fuels, production of carbon, unprecedented population growth, and mass extinction. These developments threaten the rights of humans and other-than-humans to just and sustainable lives.

In exploring ways of representing life in the Anthropocene, this work articulates innovative literary forms such as ecobiography (the representation of a human subject's entwinement with their environment), phytography (writing the lives of plants), and ethological poetics (the study of nonhuman poetic forms), providing scholars and writers with innovative tools to think and write about our strange new world. In particular, its recognition of plant life reminds us of how human lives are entwined with vegetal lives. The creative and critical essays in this book, shaped by a number of Antipodean authors, bear witness to a multitude of lives and deaths.

The chapters in this book were originally published as a special issue of *a/b: Auto/Biography Studies.*

Jessica White is based at The University of Queensland, Australia. Her memoir about deafness, *Hearing Maud*, won the 2020 Michael Crouch Award for a debut work of biography. She is currently writing an ecobiography of nineteenth-century botanist Georgiana Molloy.

Gillian Whitlock is Emeritus Professor in Communication and Arts at The University of Queensland, Australia. She is currently writing *Inhumanities: Refugee Lives in the Archives*, a life narrative of asylum seekers in detention at the Nauru camp, and the testimony of the things they left behind.

Life Writing in the Anthropocene

Edited by
Jessica White and Gillian Whitlock

Routledge
Taylor & Francis Group
LONDON AND NEW YORK

First published 2021
by Routledge
2 Park Square, Milton Park, Abingdon, Oxon OX14 4RN

and by Routledge
605 Third Avenue, New York, NY 10017

Routledge is an imprint of the Taylor & Francis Group, an informa business

British Library Cataloguing in Publication Data
A catalogue record for this book is available from the British Library

ISBN: 978-0-367-70433-9 (hbk)
ISBN: 978-0-367-70436-0 (pbk)
ISBN: 978-1-003-14631-5 (ebk)

Typeset in Minion Pro
by Newgen Publishing UK

Publisher's Note
The publisher accepts responsibility for any inconsistencies that may have arisen during the conversion of this book from journal articles to book chapters, namely the inclusion of journal terminology.

Contents

Citation Information

The chapters in this book were originally published in *a/b: Auto/Biography Studies*, volume 35, issue 1 (May 2020). When citing this material, please use the original page numbering for each article, as follows:

Introduction
Life: Writing and Rights in the Anthropocene
Jessica White and Gillian Whitlock
a/b: Auto/Biography Studies, volume 35, issue 1 (May 2020), pp. 1–12

Chapter 1
From the Miniature to the Momentous: Writing Lives through Ecobiography
Jessica White
a/b: Auto/Biography Studies, volume 35, issue 1 (May 2020), pp. 13–33

Chapter 2
Period Rhetoric, Countersignature, and the Australian Novel
Thomas Bristow
a/b: Auto/Biography Studies, volume 35, issue 1 (May 2020), pp. 35–61

Chapter 3
Writing Toward and With: Ethological Poetics and Nonhuman Lives
Stuart Cooke
a/b: Auto/Biography Studies, volume 35, issue 1 (May 2020), pp. 63–79

Chapter 4
Becoming D | other: Life as a Transmuting Device
Astrid Joutseno
a/b: Auto/Biography Studies, volume 35, issue 1 (May 2020), pp. 81–96

Chapter 5
Writing the Lives of Plants: Phytography and the Botanical Imagination
John Charles Ryan
a/b: Auto/Biography Studies, volume 35, issue 1 (May 2020), pp. 97–122

Chapter 6
"If a Tree Falls ...": Posthuman Testimony in C. D. Wright's Casting Deep Shade
Eamonn Connor
a/b: Auto/Biography Studies, volume 35, issue 1 (May 2020), pp. 123–143

Chapter 7
Writing the Lives of Fungi at the End of the World
Alexis Harley
a/b: Auto/Biography Studies, volume 35, issue 1 (May 2020), pp. 145–159

Chapter 8
Planetary Delta: Anthropocene Lives in the Blues Memoir
Parker Krieg
a/b: Auto/Biography Studies, volume 35, issue 1 (May 2020), pp. 161–181

Chapter 9
Memoir and the End of the Natural World
Tony Hughes-d'Aeth
a/b: Auto/Biography Studies, volume 35, issue 1 (May 2020), pp. 183–205

Chapter 10
"As Closely Bonded as We are:" Animalographies, Kinship, and Conflict in Ceridwen Dovey's Only the Animals *and Eva Hornung's* Dog Boy
Grace Moore
a/b: Auto/Biography Studies, volume 35, issue 1 (May 2020), pp. 207–229

Chapter 11
"Desperation for Life": Writing Death in the Anthropocene
Jessica White and Gillian Whitlock
a/b: Auto/Biography Studies, volume 35, issue 1 (May 2020), pp. 231–235

Chapter 12
Writing the Cow: Poetry, Activism, and the Texts of Meat
Jessica Holmes
a/b: Auto/Biography Studies, volume 35, issue 1 (May 2020), pp. 237–250

Chapter 13

Sheep: Voice | Complicity | Precedent
Barbara Holloway
a/b: Auto/Biography Studies, volume 35, issue 1 (May 2020), pp. 251–259

Chapter 14

A Triumphal Entry, a Stifled Cry, a Hushed Retreat
Rick De Vos
a/b: Auto/Biography Studies, volume 35, issue 1 (May 2020), pp. 261–271

Chapter 15

Her Biography: Deborah Bird Rose
Stephen Muecke
a/b: Auto/Biography Studies, volume 35, issue 1 (May 2020), pp. 273–277

Chapter 16

Artist's Statement
Anna Laurent
a/b: Auto/Biography Studies, volume 35, issue 1 (May 2020), pp. 279–280

Notes on Contributors

Thomas Bristow is the Roderick Research Fellow at James Cook University, an Honorary Research Fellow through the Australian Research Council Centre for the History of Emotions at The University of Western Australia, and a Research Fellow at the Institute of English Studies, University of London. He is also editor-in-chief of the journal *PAN* (*Philosophy Activism Nature*) and the *Environmental Literature, Culture and Media* series editor at Routledge. He is the author of *The Anthropocene Lyric: An Affective Geography of Poetry, Person, Place* (2015), co-editor of *The Cultural History of Climate Change* (2016), and author of numerous essays on the environmental humanities, history of emotions, and the Anthropocene.

Eamonn Connor is a PhD candidate in the Comparative Literature department at the University of Glasgow. His doctoral research investigates how images of the ocean liner in art and literature reflect the entangled environmental and colonial histories of the Atlantic. His background spans a diverse range of disciplines and mediums: media studies, photography, video art installations, screen and cultural studies, and philosophy.

Stuart Cooke is Senior Lecturer in Creative Writing and Literary Studies at Griffith University. His books include *Speaking the Earth's Languages: A Theory for Australian-Chilean Postcolonial Poetics* (2013) and the poetry collections *Opera* (2016) and *Lyre* (2019). In 2020, he was based at the B. R. Whiting Studio in Rome.

Rick De Vos is Adjunct Research Fellow at Curtin University, conducting research in the area of anthropogenic extinction, particularly in terms of its cultural significance and the way that extinction is articulated and practiced. He has published essays in *Knowing Animals* (2007), *Animal Death* (2013), *Extinction Studies: Stories of Time, Death, and Generations* (2017), *The Edinburgh Companion to Animal Studies* (2018), and the *Animal Studies Journal* (2014–2017).

Deborah M. Fratz is Associate Professor of English at the University of Wisconsin-Whitewater, where she teaches nineteenth-century British literature. Her research interests include disability and masculinity studies, as well as life writing.

Alexis Harley is author of *Autobiologies: Charles Darwin and the Natural History of the Self* (2015) and associate editor of *Life Writing*. She has been working for well over a decade in the field of autobiography studies, with a particular focus on conversations between life writing and natural history from the eighteenth century onwards. She is currently exploring a genealogy of how literary forms and aesthetic modes have percolated through natural history writing and continue to structure the ways contemporary Western cultures think and feel about life and ecology.

Barbara Holloway most often writes creative nonfiction in the liminal zone where agricultural and ecological worlds join. Her work has appeared in essay collections and in journals such as *Westerly*, *Southerly*, *Fusion*, and *AJE*. "Sheep: Voice | Complicity | Precedent" is part of a longer memoir|manuscript. Barbara is a Visiting Fellow in Literature, Languages, and Linguistics at the Australian National University and divides her time between Canberra and the bush block that is part of Bush Retreats for Eco-Writers (thebrewnetwork.org).

Jessica Holmes is a PhD candidate in English Language and Literature at the University of Washington, Seattle, where she also teaches in the English department. Her research focuses on the relationship between the environmental humanities and contemporary poetry. She received a Master of Fine Arts in poetry from the University of Washington (2015) and a Bachelor of Fine Arts in English from Lewis & Clark College (2011). Her poems have been published in *TRANSverse Journal* and *West Trade Review*. She was a 2019 Mellon Fellow for New Public Projects in the Humanities. She was born in Norwich, England.

Tony Hughes-d'Aeth is Director of the Westerly Centre and the Chair of English and Literary Studies at The University of Western Australia. He is the author of *Like Nothing on This Earth: A Literary History of the Wheatbelt* (2017) and *Paper Nation: The Picturesque Atlas of Australasia* (2001), and has a sustained interest in literature and the environment. He has won the Ernest Scott and W. K. Hancock prizes for Australian history, been shortlisted for the Western Australian Premier's Literary Award (Non-Fiction), and received a special commendation for the Australian University Heads of English award for the best work of literary scholarship in Australia in 2017. Tony was the co-editor of *Westerly Magazine* from 2010 to 2014.

Astrid Joutseno works at the University of Helsinki and is writing her PhD on motherhood experiences and expression on mommyblogs. Her research focuses on the intersections of digital and material subjectivities, and motherhood as an empowering position from which to do creative research and art. Joutseno is also known as Astrid Swan, a professional songwriter and performer.

Parker Krieg is Postdoctoral Research Fellow in Environmental Humanities at the University of Helsinki, affiliated with the Humanities program in the Faculty of Arts and the Helsinki Institute of Sustainability Science. He studies the environmental cultures of post-Fordism and is currently working on issues of cultural memory and environmental archives in contemporary literature. His recent articles have appeared in *Studies in American Fiction* and *Ecozon@: European Journal of Literature, Culture and Environment.*

Grace Moore teaches at the University of Otago, New Zealand. Her research interests include Victorian crime fiction, Dickens, Trollope, the history of emotions, and the environmental humanities. She is the author of *Dickens and Empire* (2004), which was shortlisted for the New South Wales Premier's Award for Literary Scholarship in 2006, and *The Victorian Novel in Context* (2012). Her most recent book is the co-edited collection *Victorian Environments* (with Michelle J. Smith, 2018). Grace's current book project is on Anthony Trollope and the environment, and she is also completing a study of settler bushfire literature.

Stephen Muecke is Professor of Creative Writing in the College of Humanities, Arts and Social Sciences at Flinders University, South Australia, and is a Fellow of the Australian Academy of the Humanities. *Recomposing the Humanities with Bruno Latour*, edited with Rita Felski, is forthcoming with Johns Hopkins University Press, and *The Children's Country: The Creation of a Goolarabooloo Future in North-West Australia*, co-authored with Paddy Roe, is contracted with Rowman and Littlefield International.

John Charles Ryan is poet and scholar with appointments as Postdoctoral Research Fellow in the School of Humanities, Arts and Social Sciences at the University of New England, Australia, and Honorary Research Fellow in the School of Humanities at The University of Western Australia. His teaching and research cross between the environmental and digital humanities. His work includes *Digital Arts: An Introduction to New Media* (2014; co-author), *The Language of Plants: Science, Philosophy, Literature* (2017; co-editor and contributor), *Plants in Contemporary Poetry: Ecocriticism and the Botanical Imagination* (2017; author), *Southeast Asian Ecocriticism: Theories, Practices, Prospects* (2017; editor and contributor), and *Forest Family: Australian Culture, Art and Trees* (2018; co-editor and contributor).

Jessica White teaches and researches at The University of Queensland. She is the author of *A Curious Intimacy* (2007), *Entitlement* (2012), and a hybrid memoir about deafness, *Hearing Maud* (2019). Her short fiction, essays, and creative nonfiction have been published in national and international journals, and she has won awards, funding, and residencies. She is currently writing an ecobiography of nineteenth-century botanist Georgiana Molloy, Western Australia's first female scientist. In 2020, she will be a Fellow at the Rachel Carson Center for Environment and Society at Ludwig-Maximilians-University, Munich, and at the Edinburgh Environmental Humanities Institute, University of Edinburgh.

Gillian Whitlock is Professor Emerita of English in the School of Communication and Arts at The University of Queensland and a Fellow of the Academy of the Humanities. She is the author of *The Intimate Empire* (2000), *Soft Weapons: Autobiography in Transit* (2006), and *Postcolonial Life Narrative: Testimonial Transactions* (2015), as well as numerous chapters and articles on life writing. Her current research focuses on the testimonial cultures of asylum-seekers on Manus and Nauru.

Laura L. Wood, PhD, is the current President of the North American Drama Therapy Association and Assistant Professor at Molloy College. Her focus and research areas include the treatment of trauma and dissociation, eating disorders, attachment, grief and loss, aphasia, and the use of drama therapy. Specifically, she is interested in how therapeutic theater can help individuals in different forms of recovery. She has both lectured nationally and internationally and published in these areas. In 2015–2016, she was awarded the Drama Therapy Fund research grant and the Molloy College Student Faculty Recognition Award.

INTRODUCTION

Life: Writing and Rights in the Anthropocene

By Jessica White and Gillian Whitlock

Figure 1. Tree. ©*Michael Leunig*

Two images locate this special issue. The first is the portrait on the cover of "Life Writing in the Anthropocene," from Anna Laurent's immersive installation *Flora: Exploration to Extinction*. This site-specific installation was commissioned for the corridors of the Exhibitionist Hotel in South Kensington, London, in 2018. Arranged on the saturated color of the emerald walls, the gilt frames emphasize the portrait presentation of black silhouettes of species of extinct or endangered flora, each accompanied by an information panel including the plant name, native ecology, and ecological pressures. Laurent chose black silhouettes to signify the absence of these plants in the landscape, as well as a historical reference to conventions of Victorian portraiture and nineteenth-century plant exploration, a period during which "species traveled the globe and luxurious images of flora were fetishized," particularly in an era of perceived ecological abundance.[1]

Here, the gilt frames are surrounded by dried seed heads and leaves, preserved relics of plant life that are spray-painted gold. The specific silhouette we have selected for the cover is the miniature cactus Biznaguita (*Mammillaria sanchez-mejoradae*), which is endemic to one location in Neuvo León, Mexico. Its population has reduced by seventy-five percent in fifteen years and it is at risk of extinction due to illegal collection and climate change.[2]

Laurent's immersive installation signals the turn to plant life (and extinction), which is a theme of this issue on the Anthropocene. The juxtaposition of portraiture—traditionally a depiction of an individual person of significance in humanist life narrative—and endangered flora in Laurent's artwork introduces the immersion in "phytographia"—"human writings about plant lives as well as plant writings about their own lives"[3]—which is a feature of this special issue. Laurent's turn to portraiture and extinction also captures a key feature of the Anthropocene: it is, "first and foremost, about death on an unimaginable scale".[4]

This special issue is a companion to the preceding issue of *a/b: Auto/Biography Studies* on "Engaging Donna Haraway: Lives in the Natureculture Web." We sustain its questioning of what constitutes a life, how that life is narrated, and what lives matter in autobiography studies now, and the turn to "phytographia" in this issue complements the focus on animal lives there. As Cynthia Huff argues in her essay on the autobiographical pact in this preceding issue, the environment of our research field has changed significantly, with a posthuman turn that eschews the human as a principal actant in favor of the distributive agency of the animal, the machine, inanimate matter, and the human.[5] For almost a decade now, special issues of journals, such as these companion issues, have led this turn from the human signature to "posthuman lives."[6] The distinctive contribution of this specific issue to this tradition of innovative scholarship is an emerging recognition of plant life, which thrives here in "ecobiography" (White), "ethological poetics" (Cooke), "phytography" (Ryan), "the witness tree" (Connor), "ethnomycology" (Harley), the commodification of plant life in the "Plantationocene" (Krieg), and "wheaten childhoods" of settler colonialism (Hughes-d'Aeth). Laurent's installation and its turn to portraiture and the silhouette to represent plant life (and extinction) captures this botanical turn, as well as the definitive force of thinking in terms of the age of the Anthropocene—the "extravagant wastage of lives and of earth."[7]

That autobiography studies as a field of research should engage with debates in environmental humanities on the issue of the Anthropocene in this way is not surprising. Kate Douglas and Ashley Barnwell argue in their recent reflections on *Research Methodologies for Auto/biography Studies* that this has been one of the most vibrant interdisciplines to emerge in the humanities and the social sciences in the past decade, moving across literary and creative writing, history, gender studies, education, sociology, and anthropology.[8] As Huff asserts, this transforms the autobiographical pact into the zoetrophic pact, encompassing "not just 'bio' but *zoe*—that is, life that has generally been considered so insignificant as to be killable."[9] These companion issues of *a/b: Auto/Biography Studies*, addressing the politics and poetics of the environment in this interdiscipline, capture the dynamism and agency of this field and its imbrication in issues of social justice, which transform the discourse of human rights and narrated lives by extending the ethics of recognition to the material presence of other-than-human life.

As a concept, the Anthropocene is distinguished by its grasp of humans as a geophysical force, shaping the planet's biophysical systems through the combustion of fossil fuels and production of carbon, unprecedented population growth, transformation of the earth's land surface and water flows, and mass extinction. While scholarship on these man-made changes to the earth has circulated since the nineteenth century,[10] atmospheric chemist and Nobel laureate Paul Crutzen and biologist Eugene Stoermer formalized the term "Anthropocene" in 2000 to "emphasize the central role of mankind in geology and ecology."[11] The term has pervaded the sciences and humanities, and is now under formal consideration as a possible addition to the Geological Time Scale by an Anthropocene Working Group of the Internal Commission on Stratigraphy, thereby joining other ages such as the Jurassic and Pleistocene.[12] The term is not without its critics. Australian novelist James Bradley resists "the way its assertion of human primacy reiterates the blindness that got us here."[13] He prefers biologist and conservationist Edward O. Wilson's term "Eremocene," the Age of Loneliness.[14] Other concepts that stake a claim here include the Capitalocene, reflecting the role of capitalism in creating the carbon bomb, as well as discussions regarding the start date for the Anthropocene, with Crutzen and Stoemer arguing for the late 1700s, a time when the effects of human activity became noticeable (ice cores, for example, record an increase of carbon dioxide), which also coincided with James Watt's invention of the steam engine, an agent of the Industrial Revolution.[15] The Necrocene points to deaths of the sixth mass extinction that surround us; the Plantationocene refers to the transformation of farms and estates into plantations that extract, via exploitation, human labor; and Donna Haraway's Chthulucene identifies an era defined by its focus on a web of relationships, "a timeplace for learning to stay with the trouble of living and dying in response-ability to a damaged earth."[16]

The second image we have chosen to introduce this special issue on life writing in the Anthropocene is a cartoon by Michael Leunig (figure 1), which takes its inspiration from Joyce Kilmer's poem "Trees." Leunig's poem is, we recognize, whimsical, although it belies a serious message. By placing trees and arboreal discourse at the center of our life-writing narratives, we question our awareness of and interaction with other-than-humans. This issue is an attempt to imagine and interrogate how to write about those "living things that have no rights," that are of value only in terms of commodification.[17] The question of human rights and narrated lives has been germane to the energy of life narrative as an interdisciplinary field. However, at any historical moment, only certain stories are tellable and intelligible to a broader audience in terms of rights discourse.[18] Questioning "the human" and, with it, some of the key tenets of the "bio" of auto|biography, such as the status of the human and the rights of other-than-human things, calls into question the ethics of recognition of rights-based discourse and its modernist language of rights and social justice for embodied individuals. This issue, like "Engaging Donna Haraway: Lives in the Natureculture Web," suggests that the posthuman turn in life-writing theory and practice now engages critically and creatively with Anthropocene landscapes of death and extinction, and "emergent and unexpected constellations of life, nonlife, and afterlife".[19]

The force of both Laurent's portraits and Leunig's cartoon suggests the importance of artistic practice as well as scholarly theory in expanding thinking on life matters and life writing. This issue of *a/b* is, as always, practice-led, beginning with a section that presents insights into the craft and construction of Anthropocene life writing. As coeditor, Jessica White's critical and creative writing has been germane to the botanical focus of this issue, and her essay on writing an ecobiography of the settler-botanist Georgiana Molloy returns to nineteenth-century botanizing, a discourse Laurent evokes in the *Flora: Exploration to Extinction* installation. Molloy has been the subject of several significant biographies to date. However, it is only in ecobiography that the specimens she collects feature as significant life forms. The Southwest Australian Floristic Region is the focus of this decolonizing ecobiography, and Molloy's numerous letters suggest how the unique and exquisite flora of this ecosystem are instrumental (rather than incidental) to her sense of self as a settler woman. Decolonizing ecobiography expands chronology to incorporate the deep time of both biological systems and Indigenous lifeways, and here, as throughout this issue, anthropologist Deborah Bird Rose's concepts of "nourishing terrains" and "country," drawing on Indigenous knowledges, are critical. Given the scale of ecological devastation and collapse, new modes of storytelling are required to convey our interconnection with, and responsibility to, things other than humans.

Rose's work on "writing into the Anthropocene" inspires two other essays on the writing process.[20] Thomas Bristow's experimental

"Sebaldian stereometry," composed in a series of embodied "Anthropocene field notes" and "countersignatures," is crafted precisely in place and time: Townsville, North Queensland, Australia, on the eve of a federal election where climate change, the proposed Adani coal mine, and the conservation of the Great Barrier Reef become critically important in the national conversation on the environment.[21] In an echo of other pieces in this special issue, Bristow's incorporation of diary, image, and descriptions of walks with his companion animal engenders a porousness in his writing, an openness and dialogue with spaces and their other-than-human inhabitants. Stereometry, Bristow argues, may be a way in which we can "delete the privatized human 'I'" by pointing to "a world where nothing is final or absolute".[22]

Stuart Cooke's essay on ethological poetics and other-than-human lives also turns to Rose's concept of "mutualism" to explore ways of articulating other-than-human language and expression, crafting forms of poetry open to the "vastly different Umwelten of their subjects".[23] Questioning how we might broaden our understanding of the interpellation of the other-than-human into human lives, Cooke draws on the concept of "ethological poetics," or "the study of nonhuman poetic forms",[24] to encourage the understanding that there is a myriad of signals in the other-than-human world. This unsettles, as mycorrhizal fungi do, assumptions accompanying the classification of plants and animals (using Western taxonomic systems) that stymie understandings of interactions between humans and other-than-humans. This creative practice sets out to "translate the expression of a particular Australasian species or landform (such as a house fly, a tea tree, or an estuary) into human language".[25]

Indeed, the presence of Rose, who died late last year, beats in this issue like a pulse, particularly her acknowledgment of Australian Aboriginal peoples' deep and abiding relationship to country. In *Nourishing Terrains: Australian Aboriginal Views of Landscape and Wilderness*, Rose explains that country is "a living entity with a yesterday, today and tomorrow, with a consciousness and a will towards life."[26] Those who do not care for or respect their ecosystem will find that it is unable to support them: "The interdependence of all life within country constitutes a hard but essential lesson—those who destroy their country ultimately destroy themselves."[27] Given that the Anthropocene is marked by mass extinction, it was perhaps inevitable that this issue on the Anthropocene might trigger extraordinary and experimental creative essays on death, on the powers of grief and mourning, or on "memoir and the end of the natural world" (Hughes-d'Aeth; this issue).

How does the obituary function present itself in Anthropocene life writing, asks Bristow, and how will it transform existing rhetorical categories and genres, such as the elegy? This question of how the Anthropocene shapes genres and conventions recurs in Astrid Joutseno's haunting thanatography on "[b]ecoming ill with what will be the cause of my time's end".[28] This feminist intersectional analysis draws on Haraway's "tentacular thinking" and Rosi Braidotti's affirmative ethics—"choosing to look for hope while acknowledging the sadness around losses in the Anthropocene"[29]—and engages with the archive of the other-than-human self to "consciously exit life".[30] Here, once again, we return to Laurent's immersive installation, which transforms the conventions of portraiture and the silhouette to narrate the extinction of plant life.

This special issue, like its companion issue "Engaging Donna Haraway: Lives in the Natureculture Web," commemorates a scholar who transforms how we understand life matters.[31] Stephen Muecke's essay, "Her Biography: Deborah Bird Rose," is a further example of creative practice in the Anthropocene, transforming the conventions of the obituary to do justice to the "plural and complicated"[32] connections of humans and other-than-humans that are alive in Bird's research. "I have to talk a little about totemism," he explains, "because she revived the concept of totemism as a shared-life concept that distributed rights and responsibilities for ecological management".[33] The interdependence of all life in country returns here. "Deborah Bird Rose became a flying-fox woman after she went to Yarralin … The flying fox thus shared her life and extended her life into the environment in a logical way—a logic of care, rights, and responsibilities".[34]

No clear, bright line establishes a boundary between The Process and Essays, critical and creative writing, in this issue. We see, for example, bibliographies of both critical and creative writing, suggesting a body of scholarship that now inspires interdisciplinary writing on the Anthropocene: Karen Barad, Anna Lowenhaupt Tsing, Jane Bennett, Donna Haraway, Bruno Latour, Deborah Bird Rose, Timothy Morton, Amitav Ghosh, Cynthia Huff, and Stephen Muecke—among others—are all participants in this critical conversation. A distinctive contribution of this special issue on "Life Writing in the Anthropocene" to current scholarship on posthuman lives is the turn to writing the lives of plants. John Charles Ryan proposes that this is a new form of life writing: "phytographia." His essay, "Writing the Lives of Plants: Phytography and the Botanical Imagination," encourages an attentiveness that seeks and recognizes, in botanical prose and poetry (and, by extension, in the real world), the individual plant amidst a mass of foliage. Engineering the form of phytography, or, as he explains, "critical posthumanist life writing about more-than-humans," "pivots on the potential of collaborating and coauthoring narratives with plants".[35] Ryan addresses an as yet largely unexamined area in the field of life writing, which has tended to dwell on the lives of humans and other-than-human animals.

Phytography entails careful attention to individual plants and their habitus, their unique forms and voices. This attentiveness features in Eamonn Connor's essay "'If a Tree Falls …': Posthuman Testimony in C. D. Wright's *Casting Deep Shade*," which contemplates a book that is both formed and informed by trees. A three-paneled object, this "bok" is made of paper from trees and its spine is like a trunk. It materializes Wright's acts of witnessing and, as Connor writes, "a subject position that exceeds her-*self*".[36] These acts create a "beech-consciousness," which draws Wright into the assemblage of things that disrupt the singular notion of a "life." Connor's essay draws on testimony and witness as a mediating discourse between the human and the other-than-human, as Wright summons a reader who is capable of bearing witness to testimony that emerges from beyond a bounded human subject and returns to the theme of extending life into the environment with a logic of care, rights, and responsibilities.

The contingencies of living things are also laid bare in Alexis Harley's essay, "Writing the Lives of Fungi at the End of the World," an examination of three life narratives of mycorrhizal fungi and their symbionts. Tracing the ascendancy of humanism in the Enlightenment, and its subsequent unraveling with the Anthropocene, Harley illuminates how the indeterminacy of fungi (they are, for example, difficult to classify with traditional taxonomy) might prompt us to consider how we are never merely human but are constituted by our relations with other-than-human selves. Fungal agency problematizes "a common implication of the term 'Anthropocene'—namely, that the relationship between humans and the rest of the world operates in one direction only, the agential human acting upon inert nature",[37] rather than being a co-constitutive relation that generates unexpected relationships. Turning to Bruno Latour's writing on living in the epoch of the Anthropocene, Harley focuses on changing notions of the subject and subjection in response to the agency of earth, matter, and nature.

Two essays on memoir in this issue turn to colonization and race and the settler-colonial foundations of Anthropocene histories in the agricultural commodification of plant life: cotton and wheat. In "Planetary Delta: Anthropocene Lives in the Blues Memoir," Parker Krieg challenges periodization of the Anthropocene that excludes colonial and plantation geographies by situating blues memoirs in an alternative framing—the Plantationocene, which foregrounds attention to "[t]he colonial combination of field and factory, and its various returns in contemporary global capitalism".[38] Here, blues memoirs of artists from the Mississippi Delta are read as a record of life writing in the Anthropocene which speaks to contemporary environmental culture and the economic and racial history of Anthropogenic environmental change. These memoirs mirror this situatedness, Krieg writes, wherein "authors' encounters with the blues connect them to a tradition that links their life narratives to broader material memories of planetary production and exchange".[39] In a different yet related context, Tony Hughes-d'Aeth's essay on the commodification of wheat and reflections on the "wheaten childhoods" of Wallace Stegner, Dorothy Hewett, and Barbara York Main question the capacity of memoir to bear witness to the event of the Anthropocene in the Americas and Australia. Drawing on Dipesh Chakrabarty's influential essay on the impact of climate change for history, and the crisis of the historical subject that falls from this, this essay features the agency of wheat as a commodity that "works its logic into the landscape relentlessly, drawing human actors (settlers) into its incessant production while dispelling other human actors (Indigenous peoples) from the lives they have lived for millennia".[40] As global agricultural commodities, both cotton and wheat are critical agents of settler-colonial histories, and decolonizing scholarship questions the capacity of memoir to recognize the presence of these other-than-human agents that bear witness to the Anthropocene.

Although there is a distinctive arboreal turn in this special issue, to phytography rather than animalography, we have argued that a focus on the Anthropocene in life writing more generally is haunted by its definitive awareness of death. To return once again to Laurent's *Flora: Exploration to Extinction*, the arts and scholarship of the Anthropocene address death and extinction, the powers of grief and mourning, and the precarity of *zoe*—life generally considered so insignificant as to be killable. Grace Moore's essay on "animalographies, kinship, and conflict" turns to the history of the emotions, and representations of animal emotions in particular. In her reading of two animalographies—Ceridwen Dovey's *Only the Animals*, a collection of ten stories featuring an animal recounting the circumstances of its death, speaking from the afterlife, and Eva Hornung's *Dog Boy*, the fictionalized biography of an abandoned child who assimilates into a dog pack—Moore considers an important ethical issue in critical debates about animalographies (and, it follows, phytography): the projection of emotions onto other-than-human life and the propensity to turn to human rights discourse to recognize and empathize with these "living things that have no rights" (to return to Leunig once more). Animals and plants are rendered expendable, property to be "managed" and "farmed," and these animalographies focus on the "vexed question of who is kin and who is food".[41] The distinctions between which animals are "livestock" and which are "companions" are historical, cultural, and (in times of scarcity) subject to change. "[T]he primacy of the so-called civilized human animal over all others signifies a failure of empathy that now threatens to undermine all forms of life in the Anthropocene".[42]

This turn to the emotions and to empathy in the Anthropocene sets the stage for the essays in The Forum, and each of these confronts the reader by focusing on the ethics of human jurisdiction over the lives of other-than-human animals, whether this be through breeding and slaughter or for sport and curiosity in colonial times. In "Writing the Cow. Poetry, Activism, and the Texts of Meat," Jessica Holmes returns to Carol J. Adams' writing on "texts of meat" and the emerging field of vegan poetics to consider the ethics of interspecies relations in the Anthropocene. What does it mean to "write" the life of a cow? Here, too, the question of empathy and "bearing witness" to the suffering of other-than-human life returns as a key issue for thinking on life writing in the Anthropocene and, in response, Holmes argues for the reparative potential and the logic of care, rights, and responsibilities of vegan poetics. Barbara Holloway also contemplates the strained relationship between the domestic animal reared as a poddy lamb in the kitchen and the flock animal sent to slaughter. Drawing on E. O. Schlunke's *Stories of the Riverina*, she considers how the subjectivity of a flock animal might be written, particularly in a way that draws attention to its exploitation. Rick De Vos's contribution to this Forum is both literary and visual, and draws this special issue back to where we begin with *Flora: Exploration to Extinction.* De Vos's essay focuses on two specific historical acts of killing: a platypus near Ravenshoe, Queensland, in September 1921, and a Yangtze River dolphin in Hunan Province in February 1914. The collection of skulls of species at risk of extinction for display in the US National Museum of Natural History by Charles McCauley Hoy presents a conundrum: fear of species extinction leads to the mission of collecting, which risks making the threat of extinction manifest. De Vos turns to animalography to translate these killings from the perspective of the animal, in juxtaposition with the account in Hoy's field notes. He also includes three figures: images of the specimens in the museum collection and the catalogue records of these skulls. This special issue begins to close, then, as it begins - immersed in images of extinction - before making a segue to Stephen Muecke's biography of Deborah Bird Rose. Muecke, reflecting on Rose's gift of storytelling, reminds us that "stories, articulated with time that passes, are a good device for teaching and learning." Stories in the Anthropocene can teach us to be aware of, and how to interact respectfully with, the other-than-humans with whom we share our world.

There is an Antipodean turn in this special issue, and we complete this work on "Life Writing in the Anthropocene"—which time and again returns to "death on an unimaginable scale[43]—at a place and time where life writing and criticism is intimately connected to everyday life, as it so often is. Fire surges on an unprecedented scale here in the summer of 2019–2020; the term "catastrophic" has never been used so often. Rainforests burn; eucalypts explode and release pungent flammable fumes.

Embers fly many kilometers ahead of the fire fronts, which consume human, other-than-human animal, and plant life. Also catastrophic and historical is the drought-induced mass death of other-than-human life. In the summer of 2018–2019, up to a million native Australian fish, including Murray cod decades of years old, were killed in three separate incidents in the Darling River, a major waterway stretching across the state of New South Wales. We see "fish kill"—thousands of silvery bodies floating on the surface of a river—and listen to descriptions of the choking smell of rotting fish. The magnitude of these events is extraordinary. In times like these, where the logic of commodification and denial rules, the relevance of scholarly work in the humanities is subject to question.[44] Here, in this special issue, critical and creative writing and artworks on life narrative rise to the challenge of the Anthropocene in theory and practice. We are drawn into corridors where portraiture, ecobiography, phytography, obituary, stereometry, thanatography, testimony, and animalography bear witness to a multitude of lives and deaths.

The University of Queensland

Notes

1. Laurent, *Flora.*
2. Laurent, *Flora.*
3. Ryan, "Writing the Lives of Plants," 97.
4. Cooke, "Writing Toward and With," 63.
5. Huff, "Autobiographical Pact."
6. Whitlock and Couser, "Post-ing Lives."
7. Rose, "Slowly," 2.
8. Douglas and Barnwell, *Research Methodologies*, 32.
9. Huff, "Autobiographical Pact," 446.
10. Steffen et al., "The Anthropocene."
11. Crutzen and Stoermer, "The 'Anthropocene,'" 17.
12. Zalasiewicz, Williams, and Waters, "Anthropocene," 15.
13. Bradley, "Writing on the Precipice."
14. Wilson, "Beware."
15. Crutzen and Stoermer, "The 'Anthropocene,'" 17.
16. Haraway, cited in Huff, "Autobiographical Pact," 445.
17. Leunig, "Tree."
18. Smith and Schaffer, *Human Rights*, 32.
19. Nils Bubandt, cited in Joutseno, "Becoming D|other," 86.
20. See Rose, "Slowly."
21. Bristow's essay is also in memory of Heather Kerr, a researcher in the "History of the Emotions" project with a particular interest in botanical life. The influence of this major Australian Research Council project is also acknowledged in Moore's essay on animalography in this issue.
22. Bristow, "Period Rhetoric, Countersignature, and the Australian Novel," 55.

23. Cooke, "Writing Toward and With," 75.
24. Cooke, "Writing Toward and With," 64.
25. Cooke, "Writing Toward and With," 72.
26. Rose, *Nourishing Terrains*, 7.
27. Rose, *Nourishing Terrains*, 10.
28. Joutseno, "Becoming D|other," 82.
29. Joutseno, "Becoming D|other," 83.
30. Joutseno, "Becoming D|other," 86.
31. In the preceding issue of *a/b: Auto/Biography Studies* on "Engaging Donna Haraway: Lives in the Natureculture Web," Margaretta Jolly's essay "Survival Writing: Autobiography versus Primatology in the Conservation Diaries of Alison Jolly" is written in tribute to her mother.
32. Muecke, "Her Biography: Deborah Bird Rose," 274.
33. Muecke, "Her Biography: Deborah Bird Rose," 274.
34. Muecke, "Her Biography: Deborah Bird Rose," 274.
35. Ryan, "Writing the Lives of Plants," 101.
36. Connor, "'If a Tree Falls … '," 124.
37. Harley, "Writing the Lives of Fungi," 146.
38. Krieg, "Planetary Delta," 162.
39. Krieg, "Planetary Delta," 166.
40. Hughes-d'Aeth, "Memoir and the End of the Natural World," 201.
41. Moore, "'As Closely Bonded as We Are'," 214.
42. Barbara Creed, cited in Moore, "'As Closely Bonded as We Are'," 223.
43. Cooke, "Writing Toward and With," 63.
44. Attenborough, "It's Extraordinary."

Acknowledgments

We gratefully acknowledge the generosity of both Anna Laurent and Michael Leunig in allowing us to reproduce their images, which trigger our reflections in this special issue.

Disclosure Statement

No potential conflict of interest was reported by the author(s).

ORCID

Jessica White http://orcid.org/0000-0002-9291-5712

Works Cited

Attenborough, David. "It's Extraordinary Climate Deniers Remain in Australian Politics." *The Guardian*, July 7, 2019. https://www.theguardian.com/global/video/2019/jul/10/david-attenborough-its-extraordinary-climate-deniers-remain-in-australian-politics-video

Bradley, James. "Writing on the Precipice." *Sydney Review of Books*, February 21, 2017. https://sydneyreviewofbooks.com/writing-on-the-precipice-climate-change

Bristow, Thomas. "Period Rhetoric, Countersignature, and the Australian Novel." *a/b: Auto/Biography Studies* 35, no. 1 (2020): 35–62.

Cooke, Stuart. "Writing Toward and With: Ethological Poetics and Nonhuman Lives." *a/b: Auto/Biography Studies* 35, no. 1 (2020): 63–80.

Connor, Eamon. "'If a Tree Falls …' : Posthuman Testimony in C. D. Wright's Casting Deep Shade." *a/b: Auto/Biography Studies* 35, no. 1 (2020): 123–144.

Crutzen, Paul J., and Eugene F. Stoermer. "The 'Anthropocene.'" *IGBP Newsletter*, no. 41 (2000): 17–18.

Douglas, Kate, and Ashley Barnwell, eds. *Research Methodologies for Auto/biography Studies*. New York, NY: Routledge, 2019.

Harley, Alexis. "Writing the Lives of Fungi at the End of the World." *a/b: Auto/Biography Studies* 35, no. 1 (2020): 145–160.

Huff, Cynthia. "From the Autobiographical Pact to the Zoetrophic Pack." *a/b: Auto/Biography Studies* 34, no. 3 (2019): 445–460.

Hughes-d'Aeth. "Memoir and the End of the Natural World." *a/b: Auto/Biography Studies* 35, no. 1 (2020): 183–206.

Jousteno, Astrid. "Becoming Djother: Life as a Transmuting Device." *a/b: Auto/Biography Studies* 35, no. 1 (2020): 81–96.

Kilmer, Joyce. "Trees." *Poetry* 2, no. 5 (1913): 160.

Krieg, Parker. "Planetary Delta: Anthropocene Lives in the Blues Memoir." *a/b: Auto/Biography Studies* 35, no. 1 (2020): 161–182.

Laurent, Anna. *Flora: Exploration to Extinction*. https://www.annalaurent.com/flora-installation

Leunig, Michael. "Tree." March 8, 2014. https://www.leunig.com.au/works/recent-cartoons/463-tree

Moore, Grace. "'As Closely Bonded as We are': Animalographies, Kinship, and Conflict in Ceridwen Dovey's Only the Animals and Eva Hornung's Dog Boy." *a/b: Auto/Biography Studies* 35, no. 1 (2020): 207–230.

Muecke, Stephen. "Her Biography: Deborah Bird Rose." *a/b: Auto/Biography Studies* 35, no. 1 (2020): 273–278.

Ryan, John Charles. "Writing the Lives of Plants: Phytography and the Botanical Imagination." *a/b: Auto/Biography Studies* 35, no. 1 (2020): 97–122.

Rose, Deborah Bird. *Nourishing Terrains: Australian Aboriginal Views of Landscape and Wilderness*. Canberra, NSW: Australian Heritage Commission, 1996.

—— "Slowly—Writing into the Anthropocene." *TEXT*, no. 20 (2013): 1–14.

Smith, Sidonie, and Kay Schaffer. *Human Rights and Narrated Lives: The Ethics of Recognition*. New York, NY: Palgrave Macmillan, 2004.

Steffen, Will, Jacques Grinevald, Paul Crutzen, and John McNeill. "The Anthropocene: Conceptual and Historical Perspectives." *Philosophical Transactions of the Royal Society A* 369, no. 1938 (2011): 842–867.

Whitlock, Gillian, and G. Thomas Couser. "Post-ing Lives." *Biography* 35, no. 1 (2012): v–xvi.

Wilson, Edward O. "Beware the Age of Loneliness." *The Economist*, November 18, 2013. http://www.economist.com/news/21589083-man-must-domore-preserve-rest-life-earth-warns-edward-owilson-professor-emeritus

Zalasiewicz, Jan, Mark Williams, and Colin Neil Waters. "Anthropocene." In *Keywords for Environmental Studies*, edited by Joni Adamson, William A. Gleason, and David N. Pellow, 14–16. New York, NY: New York UP, 2016.

From the Miniature to the Momentous: Writing Lives through Ecobiography

By Jessica White

ABSTRACT

This article contemplates ecobiography, a little-researched form of life writing which depicts how human selves are supported and shaped by their environment. It details the author's ecobiography of botanist Georgiana Molloy (1805–1843) and the plants she collected from the Southwest Australian Floristic Region, alongside an analysis of an Australian ecobiography, Kim Mahood's *Position Doubtful*.

Life-Support Systems

The Southwest Australian Floristic Region (SWAFR) stretches for roughly 300,000 kilometers across the southwestern corner of Western Australia. Its landscape is old and weathered, persisting for millions of years because it has not been disrupted by earthquakes or volcanoes. Without disturbance or replenishment through geographical upheaval, the soil has become poor in nutrients, yet it has given rise to an incredible diversity of plant forms.[1] In order to ensure survival in this unique region, very specific relationships have arisen, such as that between dwarf hammer orchids (*Drakaea micrantha*) and their insect pollinators, thynnine wasps. These orchids release plumes of pheromones that trick a male wasp into thinking the plant is female. The wasp tracks the plume in a zigzagging flight, then lands on the flower and attempts to copulate with it. Coated in pollen, it flies to another flower and transmits the pollen, thereby fertilizing the plant.

These wasps and orchids demonstrate, in a vivid and beautiful way, the interdependence in an ecosystem, a "biological system composed of all the organisms found in a particular environment, interacting with it and with each other."[2] More than two centuries of European colonization have drastically impacted on the SWAFR's ecosystems, particularly those that support orchids, with more of these species being affected than any other plant family.[3] In order to protect this unique region and its other-than-human inhabitants such as *Drakaea micrantha*, scientists have advocated for its recognition as a biodiversity hotspot, an area in which high percentages of plant life, found nowhere else on earth, are at exceptional risk of extinction.[4] The SWAFR was declared such a hotspot in 2000.

While scientists are undertaking crucial research on the importance of the SWAFR, writers and scholars can also create and promote new ways of thinking about ecosystems and our place within them, rather than assuming, as anthropocentrism has maintained, that ecosystems are valuable insofar as they only serve humans. One such way is through transforming autobiography—the stories we tell about our selves—in ways that rethink anthropocentric approaches to human and other-than-human lives.

A number of scholars have written with intelligence and verve on the shared materiality of human and other-than-human selves,[5] but given the context of my research, I have opened this essay with a local example of the interdependence of self and ecosystem to illuminate my approach to my ecobiography of nineteenth-century botanist Georgiana Molloy and the SWAFR in which she collected. Although *Drakaea micrantha* is not listed among her specimens,[6] the remarkable relationships between pollinators and their flowers were responsible for the creation of flora that radically shaped her sense of self. Over the past two decades, while crafting a story of her life and the other-than-human lives with which she interacted, this wasp and orchid have prompted me to rethink the origins of autobiography.[7]

Sidonie Smith and Julia Watson open *Reading Autobiography* with a reference to the Greek etymology of the term "autobiography"—*autos* denotes "self," *bios* denotes "life," and *graphe* denotes "writing"—providing a neat definition of "self life writing."[8] The term emerged in the West in the late eighteenth and early nineteenth centuries, and analyses of the form have historically been "drawn seamlessly into supporting the belief and values of an essentialist or Romantic notion of selfhood," in which a unified selfhood is seen as representative of universal human nature.[9] This interpretation of autobiography is one that is emphatically, definitively anthropocentric, with human life at its core.

However, scholarship on posthumanism has destabilized the concept of a universal subject and queried where the self begins and ends. As the introduction to a special issue of *Biography* on posthuman lives indicates, the critical approaches of "feminism, postcolonialism, poststructuralism, queer theory, postmodernism, disability studies—all have influenced recent criticism by raising questions about the status of the human subject and subjectivity."[10] The effect of this, posthumanist scholar Rosi Braidotti notes, is that it compels us to question "what exactly is the basic unit of common reference for our species, our polity and our relationships to the other inhabitants of this planet."[11] These cultural shifts suggest that the human is not—and never has been—a unified, isolated subject. If we cannot have a life without the lives of other inhabitants such as fungi, bacteria, air, or orchids, then it becomes desirable, in creating an autobiography, to include the lives that sustain and shape it: the *autos* of a biography should include our environment. This is essential to the art of ecobiography and, in the case of my own work in progress, to the recognition of plant life, particularly that of the SWAFR.

"My prevailing passion for Flowers"

My apprenticeship in ecobiography began at the age of twenty-one, when I traveled overseas for the first time on my own to the University of California, Berkeley, for a year-long exchange. Despite California's blue skies and bright days, I missed the dry, hot Australian summer, the piquant smell of eucalyptus after a storm, and starbursts of wattle against olive-green hills in spring. My homesickness propelled me to the Australian literature section of the library, where I randomly pulled a book from the shelf. Patterned with dark-red leaves, it was titled *An All Consuming Passion: Origins, Modernity, and the Australian Life of Georgiana Molloy.*[12] Intrigued, I sat on the floor, my back against the bookshelf, and became acquainted with its subject, Georgiana Kennedy.

Born in 1805 to a wealthy family, Georgiana spent her childhood at Crosby Lodge, an estate near Carlisle in England, learning about plants and gardening within its walled garden. When she was fifteen, her father, who had been made Deputy Lieutenant of Cumberland, was killed by a fall from his horse. The family moved to Rugby for the sons' education, but due to friction with her mother, Georgiana preferred to spend her time with her friends the Dunlops, who lived near Glasgow. At twenty-four, with her options narrowing, Georgiana married John Molloy, a soldier who fought in the Napoleonic Wars, and agreed to sail with him to the fledgling colony of Swan River (now Perth) in southwest Western Australia. By the time the Molloys arrived in Swan River in 1830, Georgiana was heavily pregnant. They found that all the land in the area had been allocated by the government, so they sailed south to Augusta, where they established the first British colony. Two weeks after their arrival, Georgiana bore her first child, but the baby lived only ten days.

As I read Georgiana's account of her daughter's decline, and her subsequent comment in a letter to her mother that, five days after the baby's burial, she planted white lilies and Cape bulbs "to dispel the sad blank her death occasioned,"[13] I thought of my own mother, an emigrant from New Zealand to Australia, who lost her son when he was eight months old. My parents buried him beneath a stand of pine trees on our farm. Like Molloy, my mother is an avid gardener and when she arrived at the farm, friendless but for my father, she transformed the land surrounding our worker's cottage into flower gardens, an orchard, and vegetable beds. Chlorophyll was my heritage, and I recognized Molloy's industrious planting as a way of grounding herself in a new location. After six months, her garden was yielding daily vegetables and salads for dinner, and this, she boasted, "is more than anyone can say, not even the Governor."[14]

Molloy bore another three daughters and a son, but had little domestic help in what, to her, were rough, lonely, and exhausting conditions. In 1836, when he was nineteen months old, her son fell into a well and drowned. This prompted what Molloy described as a "dangerous illness"—most likely a nervous breakdown.[15]

Just prior to this event, Molloy received a box of English seeds and a letter from Captain James Mangles. Born in 1786, Mangles was the nephew of the director of the British East India Company and the principal of F. & C. F. Mangles, which sent convict ships to Sydney. He joined the navy when he was young, and had reached the position of officer by the time he left in 1815. He was elected a Fellow of the Royal Society in 1825 and, in 1830, he was co-founder and one of the first Fellows and members of the council of the Royal Geographical Society. He had connections with John Lindley, the first Professor of Botany at University College, London; with the Loddiges nurserymen of Hackney; and with Joseph Paxton, a gardener at Chatsworth and the designer of the Crystal Palace for the 1851 Great Exhibition. Returning from travels in Syria and Egypt, Mangles met James Stirling, a fellow Royal Navy officer. Back in England, he invited Stirling to his uncle's estate, where Stirling met Ellen, Mangles' cousin. Stirling was smitten and, when Ellen came of age, they married and sailed to Western Australia. Stirling, appointed governor of the colony, was responsible for managing its establishment. In 1831, Mangles sailed to Perth to visit his friend and cousin, and through them he made contact with Molloy, who was an acquaintance of Ellen.

Molloy's familiarity with collecting stemmed from her education as a young woman of leisure. Botany was a polite activity in the lives of English girls and women of the late eighteenth and early nineteenth centuries, being "both a fashionable form of leisure and … an intellectual pursuit."[16] She had sailed to Australia with intentions of collecting, for she packed a *hortus siccus*, a book into which pressed and dried specimens were fastened for identification purposes. However, she wrote in her first letter to Mangles, dated March 21, 1837, that her time was "so much infringed on" that she had not had a chance to use it.[17]

After the death of her son, Molloy needed a diversion from her grief. Drawing on her knowledge of collecting flowers, she went into the bush to find seeds for Mangles, explaining that "[s]ince my dear Boy's death I have up to the present time daily employed myself in your service."[18] As she gathered plants and seeds, her perception of her surroundings began to shift. Where, two years after her arrival, she described the bush to her sister as "the unbounded limits of thickly clothed dark green forests where nothing can be described to feast the imagination,"[19] now she focused on individual flowers and came to adore them. She wrote to Mangles on June 30, 1840, ten years after she arrived in the country and three years after she began collecting for him: "you ought to be interested in the *Isopogon*, and *Petrophila* [*sic*];—they are beautiful beyond description, such flowers of imagination; I am now in raptures when I think on them, in searching to come suddenly on such gems, and be surrounded by them, makes you for a time think you are in Fairy Land."[20] The contrast between these two descriptions—a place which fails to ignite Molloy's fancy as opposed to the flowers that had put her in "raptures," and the indistinguishable mass of vegetation which differs from the specificity of "*Isopogon*" and "*Petrophila*"—signals the dramatic shift in Molloy's awareness of her environment.

I recognized, from Molloy's clear delight in writing about flowers and her desire to be "free from domestic drudgery,"[21] how the bush was a solace for her, as it had been for me. As an unhappy adolescent, I would pull on a pair of joggers and run through the hills at the end of each school day. The heat that wrapped around my body, the warbling magpies, and the thick thrum of cicadas were a diversion from my struggles to fit in at school. In the quietness of the bush, there were no demands on me other than the propulsion of my body over rocks and spinifex.

In the library, I straightened my spine against the bookcase and imagined Molloy's strong and nimble fingers breaking the stem of a *Darwinia citriodora*. She peers closely at its yellow and red flowers, the piquant lemon scent scraping the insides of her nostrils. Electrified, I closed the book and sprang from the floor, hobbled until the pins and needles in my feet disappeared, then strode to the loans counter to borrow it.

When I returned to Australia at the end of that year, I discovered an earlier biography, Alexandra Hasluck's *Portrait with Background*,[22] which contained longer extracts of Molloy's writing. Spanning several pages, these letters used language, character, and dialogue to create dramatic scenes. Yet neither Lines nor Hasluck commented on the craft of

Molloy's writing, nor the evident shift in her life once she began to collect seeds and specimens, nor the quiet desperation in one of her last letters to Mangles when she wrote, "I shall be at a loss to know what to send you of novelties in coming season I believe I have sent you every thing worth sending *far* and *near*, therefore what am I to do."[23] Without a reason to collect flowers, she would lose the opportunity to remove herself from what she described, in a long letter of January 28, 1838, as the "manifold duties of domestic drudgery"[24] and engage with her environment. Her lines echoed the unsettlement I had experienced in the US at being so far away from Australia, as well as the sense that I had left part of my self behind.

Molloy's story stayed with me for fifteen years while I completed other writing projects. I located the originals of her letters in archives in Perth, Australia and Carlisle, England, and found a stark contrast in tone between the letters she wrote to her family and those she composed to Mangles. Where she wrote to her mother about the birth and development of her children, to Mangles she asserted, "I never met with any one who so perfectly called forth and could sympathize with me in my prevailing passion for Flowers."[25] The phrase "called forth" suggests that her obsession with plants had lain latent until Mangles activated it with his request for seeds.

During this time, another biography was published, Bernice Barry's *Georgiana Molloy: The Mind That Shines*. While Barry's research is far more detailed than Hasluck's or Lines', it remains another conventional cradle-to-grave biography, dwelling more on Molloy and her husband's lives in Britain before they met and emigrated to Australia. It was clear that Molloy depended on the SWAFR's flora for her well-being, but the three autobiographies did not convey its significance to her sense of self. I realized that I needed to write not just an autobiography of a person or of an environment, but something which detailed the imbrication of Molloy's self with her environment, and which accorded weight to the plants that shaped her. I reached out for a word to grasp this synergy of self and environment: *ecobiography*.

Ego and Eco

The term "ecobiography" first appeared in Cecilia Konchar Farr and Philip Snyder's 1996 essay "From Walden Pond to the Great Salt Lake: Ecobiography and Engendered Species Acts in *Walden* and *Refuge*." The authors describe ecobiography as "a life-story constructed according to a pattern divined internally through the Self's interaction with the external environment, especially Nature, the multiple exchanges of which (re)present a kind of ecosystem of the Self."[26] In these life stories, they continue, "it is impossible to tell where the Self ends and Nature begins or where Nature ends and the Self begins: ego and eco are inextricably intertwined."[27] In 2001, Smith and Watson included a definition of ecobiography in *Reading Autobiography*, describing it as "life writing that interweaves the story of a protagonist with the story of the fortunes, conditions, geography, and ecology of a region and reflects on their connection (sometimes failed) as a significant feature of the writing."[28] In 2010, they expanded and reprinted their volume, providing a wider range of examples of ecobiography. However, in the two decades since the term's provenance, there has been remarkably little critical attention to the form and what it constitutes, perhaps reflecting the entrenched anthropocentrism of life writing. Of the criticism that has been published on ecobiography to date, there are two major themes: the dissolution of boundaries between human and other-than-human, and attention to local environments and ecosystems.[29]

In an Australian context, this means heeding Indigenous Australian peoples' relationship with place, as they are the continent's traditional custodians and have cared for it for 60,000 years.[30] This long history has created a deep and abiding relationship between Indigenous peoples and the human and other-than-human inhabitants of these countries, as ethnographer Deborah Bird Rose describes in *Nourishing Terrains*: "A 'healthy' or 'good' country is one in which all the elements do their work. They all nourish each other because there is no site, no position, from which the interest of one can be disengaged from the interests of others in the long term. Self-interest and the interest of all of the other living components of country (the self-interest of kangaroos, barramundi, eels and so on) cannot exist independently of each other in the long term."[31] This understanding of interconnectedness accords with the definition of "ecosystem" mentioned earlier. It demonstrates Indigenous Australian peoples' understanding of their interactions and interdependencies with the other-than-human world.[32]

Crucially, this is a world which is animate. In an essay on the life and work of Australian philosopher, activist, and writer Val Plumwood, Rose describes how Plumwood "understood that Aboriginal Australians always live within a world that is buzzing with multitudes of sentient beings, only a very few of whom are human."[33] Lardil man Goobalathaldin Dick Roughsey describes this sentience, and human interaction with it, in *Moon and Rainbow*, the first autobiography published by an Indigenous Australian in English: "Buldingu, (Venus) the woman star, is first to go down in the evening. She goes to make a big fire to cook the fish which the men stars, the hunters, will be bringing home later. These men stars,

Bidgingu and Gidbu Bulan, come up at ten o'clock or later. They carry on their backs Yule, the Milky Way. Yule is a road or a path for the hunters to travel on during the night. … When the old men saw Gidbu Bulan coming up, one of them would go down and look at the water."[34] The Lardil people's recognition of the stars' agency is evident in the description of Buldingu's creation of fire, while the interaction between human and other-than-human is indicated by the path which the hunters travel along, created for them by Bidgingu and Gidbu Bulan. The hunters' attentiveness to the animism of other-than-human beings is indicated by the ways in which they take their cues from these beings.

Critical engagement with Australian Indigenous life writing tends to emphasize identity formation and trauma effected by colonization and the Stolen Generations. As Michèle Grossman writes in her survey of Indigenous autobiography, many of these texts are "political interventions in the most fundamental of senses,"[35] counternarratives that highlight the impact of invasion and government control on the lives of Australia's Indigenous peoples. However, reading life writing by Indigenous writers, such as Roughsey's *Moon and Rainbow*, Kim Scott's *Kayang and Me*, or Oodgeroo Noonuccal's *Stradbroke Dreamtime*, has another important function: it reminds non-Indigenous readers of how Indigenous peoples have interacted with ecosystems over thousands of years in ways which ensure the balance and survival of themselves and of other-than-humans.[36]

It was clear to me that, as Molloy was influenced by an environment which had been shaped by its traditional custodians, the Noongar peoples,[37] and she was complicit in the dispossession of those peoples through occupation of their land and the colonial trade in plants (which exerted linguistic control through classification and renaming),[38] it was necessary to include a discussion of the relationship between Noongar peoples and their environment in my ecobiography, as well as how this has been impacted by colonization. It was therefore important to locate examples of ecobiography by non-Indigenous Australians which illuminated the ways in which settler-colonial selves were shaped by their environments, while at the same time remaining attentive to the history of Indigenous Australians' custodianship of those environments.

Australian Ecobiography

Critical commentary on ecobiography in Australia, where a tradition of nature writing emerged later than in the US and Europe, continues to develop.[39] Australian scholar Inga Simpson argues that Australia's nature writing is shaped by its engagement with Australia's unique flora and fauna and by Indigenous Australian storytelling, which closely observes and is bound up with the natural world.[40] These observations can also be applied to Australian ecobiography, one example of which is Kim Mahood's *Position Doubtful: Mapping Landscapes and Memories.*[41]

In her late childhood, Mahood's family moved to what they named Mongrel Downs, a cattle station situated some seven hundred kilometers northwest of Alice Springs in the Tanami Desert. In 1992, the station was returned to its traditional owners and renamed Tanami Downs. When Mahood's father died in a helicopter crash, she made a pilgrimage back to this country, describing it in her first memoir, *Craft for a Dry Lake.* This trip morphed into annual journeys, which included artist field trips, visits to Mahood's former family home on the cattle station, volunteer work in the Indigenous art center at Balgo, an artist residency at the Tanami mine, and map-making expeditions and environmental projects with the Walmajarri custodians of Lake Gregory.[42] Mahood describes these journeys in her second book, *Position Doubtful,* focusing on how her years spent with Walmajarri people as a girl and, later, as a woman shaped the way she perceives country and herself. Her ecobiography is "a kind of love story … for a remote pocket of inland Australia."[43] The desert, she explains, "took up residence in my psyche when I was very young, in an impersonal, obdurate sort of way that made it impossible for me to form lasting attachments to any other place."[44] It also transformed her perspectives and visual practice.

Mahood opens *Position Doubtful* with a behest to the reader to imagine that "the document you have before you is not a book but a map. It is well-used, creased, and folded so that when you open it, no matter how carefully, something tears and a line that is neither latitude nor longitude opens in the hidden geography of the place you are about to enter."[45] This passage, with its reference to a "hidden geography," the "tearing" of conventional coordinates, and the creation of a line that does not conform to conventional coordinates, signals that this will be a text which explores Mahood's shifting relationship with the desert.

Mahood uses art and writing to express what she describes as "the improvised life I have chosen to live for the past twenty years, straddling two worlds, settling in neither."[46] One world is in the city and the other is in the "remote, predominantly Aboriginal, barely sustainable, thinly populated pocket of desert."[47] When she was a baby, Mahood acquired the skin name of "Napurrula," which means that she belongs "to the skin group with traditional links to Tanami Downs."[48] When she returns to the desert on her repeated trips, she tries to negotiate her relationships with those such as Patricia, who is her daughter in the kinship system, and comments, "Between us hovers something that is not quite friendship, which stems from an awareness of the ambivalent ground we occupy. Our shared connection to country carries with it some form of mutual obligation, the nature of which I am uncertain."[49] Mahood's position is indeed "doubtful."

As such, it is apt that she arrives at the motif of the grid in her work, not only in an attempt to ground and orient herself, but also to explore the way in which she perceives space and time in the desert, and how this is inflected by her interactions with Walmajarri people. In 1993, when she stood before Indigenous paintings from the Holmes à Court collection at Vasse Felix in southwest Western Australia, Mahood saw for the first time how her country was "represented in a way that spoke directly of its essence."[50] At the same time as Mahood recognized that "here was an iconography and an organising principle that perfectly expressed my experience of the country in which I had grown up," she knew that she could not use it.[51] Mahood is not Indigenous and "the ancestral imprint that gave the desert paintings such authority and power was not available to me."[52] Instead, she needed to find an equivalent in her own culture, and this led her to two critical elements of the Western visual tradition: the grid and the horizon line. These are used to represent three dimensions on a two-dimensional surface, "producing the perspective we take for granted in traditional western painting."[53] Meditating on the motif of the grid, Mahood writes, "The grid has had a bad press as a colonial tool, especially when applied to mapping, but that is only part of the story. The mapping grid that is based on latitude and longitude distorts the shape of continents when it is used as the basis for a two-dimensional map. But the co-ordinates it represents, north/south and east/west, are embedded in the natural world: the rising and setting of the sun and moon; the axis on which the planet spins."[54] Her receptiveness to the alternative capacities of the grid, particularly in terms of its connection to the natural world, means that she is open to modes of seeing alternative to those inherent in the Western visual arts tradition.

One of the projects with which Mahood became involved was the Canning Stock Route Art Project. Conceived in 2006 by Tim Acker, the project aims to tell the story of the Canning Stock Route through Indigenous Australian voices. The route, between Halls Creek and Wiluna, was established in the early twentieth century to move cattle down from the north of Western Australia, but it was only used with frequency from the 1930s to the 1950s. The Paruku artists paint their responses to their country, the section of the route that "falls under the jurisdiction of the eastern Walmajarri."[55] Watching them is like, Mahood writes, "a force field in which the gestures of laying paint on canvas completes a circuit that begins and ends with country, the body of the painter serving as a transformer to direct some of that perpetual current into a visible expression of the charge it carries."[56] The Paruku artists' "maps" are not a means of inscribing the land in terms of its utility as capital. Rather, they are a writing of the country with art through themselves and of themselves.

One of the artists is Billy Patch, of the Manyjilyjarra language group and Milangka skin group. He hand-drew a grid in the sand to "represent the songlines—establishing them as the fundamental co-ordinates of the desert—and superimposed the stock route on the grid like the transient track of a snake."[57] Patch used a European tool—the grid—to explain how "in the eyes of Aboriginal people, the Canning Stock Route cut through a pre-existing network of cultural and ceremonial boundaries, songlines and laws. Each intersecting point on this grid represents a different water source. The lines connecting them symbolise the songs that are sung for these sites."[58] In this instance, the grid is both a symbol of colonization and a connection to the natural world, being used to represent the vital points of the area's ecosystem. Mahood observes that this "doubleness was where the charge lay."[59] The grid is not just a tool for mapping oppression but a means to explain a piece of critical knowledge: the availability of water sources in a desert. By viewing Patch's interpretation of the grid, Mahood understands both her colonial history and a new way of perceiving her world.

The "charge" that Mahood experiences runs through her body at other points in the text. When she is exposed to the art of Indigenous painters in Balgo in 2001, their "embodied representations of country threaten to overwhelm whatever tentative intuitions I've been trying to formulate, teasing out the root of the difference between how I've been taught to interpret the world, and this other multi-layered response to the physicality of place."[60] This sense of destabilization is not only caused by visual sensations; it is also aural. When Mahood hears the cry of a stone curlew, which "seems to come from some other dimension," she is shaken loose "from my own filtered consciousness into the vibrating strangeness of a world in which there are none of the contemporary reference points I take for granted."[61] These moments "offer a glimpse of some other kind of framework for knowing this country, this place we live in and belong to."[62] Such unsettlement is crucial to the dismantling of the notion of the unified Western subject. When she steps into "strangeness," Mahood becomes receptive to the animated life of the country. Her lessons from Indigenous Australian art and its creators help her to understand and chronicle the indivisibility of the human and other-than-human: "To learn the names of trees and grasses, the times of their seeding and flowering, the glimpse they offer into the grand slow cycles of nature, is to see your own life written there, and passing. To know the geography of a place is to know why we have always made stories in which our own human stuff is indivisible from the stones and creeks and hills and growing things."[63] In this passage, which can be read as a definition of ecobiography, the two main themes of the genre become apparent: human life is placed firmly within an ecosystem, rather than suggesting that it is apart from it, so that the distinction between human and other-than human is blurred. The role of storytelling is also highlighted: the writer not only narrates a particular local place, but is narrated by it, changing one's self and the stories which that self creates.

From the Miniature to the Momentous

The impact of generations of Indigenous Australian life-narrative and key ecobiographical texts by Mahood, among others, has generated a context for my ecobiography in progress on Molloy, "From the Miniature to the Momentous." But where Mahood uses maps and the grid to position herself in her story and explore alternative modes of seeing and shaping that story, I use the plants of the SWAFR.

Many of the flowers in the SWAFR are no bigger than a fingernail. Molloy, who lamented in her letter to Mangles that "I long to see again a large flower," needed to observe and distinguish between the tiny flowers, collect their seeds, and mount the dried specimens into the *hortus siccus*.[64] Often she took her daughters with her to collect plants as, she wrote to Mangles in this same letter, "their eyes being so much nearer the ground they have been able to detect many minute specimens and seeds I cou'd not observe."[65] Yet these plants, tiny as they are, ignited Molloy's passion, to the point where she wrote to Mangles, "I discovered a plant I have been almost panting for, a very small neat white blossom, on a furze looking Bush."[66]

For Georgiana, the pleasure of collecting lay not only in the aesthetics of the plants she gathered, but also in the opportunity it offered to write about the flowers she encountered. In a letter to Mangles of March 14, 1840, she wrote, "I beheld a Tree of great beauty. … the flowers gave the character not only to their possessor but to the surrounding locality, the flowers are of the finest white, and fall in long tresses from the stem, some of its pendulous blossoms, are from three, to five, fingers in length, and these wave in the breeze like Snow wreaths; they are of such a downy white feathery appearance, and emit a most delicious perfume, resembling the bitter almond, and like all human or rather mortal delicacies, how quick these lovely flowers fall from the stalk on being collected."[67] Molloy enlarges on the beauty of the tree by stressing its color as "fine," its blossoms as "downy" and "feathery," and its scent as "delicious." Her use of "Snow wreath," a heap of snow blown together by the wind, emphasizes the delicacy of the blossoms. This is a particularly sensory description, indicating how much Molloy enjoyed being outside, inspecting, touching, and smelling the plants. It also shows her pleasure in her perception of the bush, commenting favorably on the tree in its "locality." At the close of her description to Mangles, Molloy writes, "on nearer view, [I] found the buds much more beautiful than the full blown flower—I regret they have assumed a yellow hue, but are lovely and elegant even in Death. The Native name is 'Danja'—and I rather think it will turn out to be a Hakea."[68]

Due to her lack of formal education in botany and the newness (to European eyes) of the flora in the SWAFR, Molloy did not identify the plant correctly at this point (in a later letter, having learned the name from a visiting botanist, she does). The plant was not a hakea, although it does belong to the *Proteaceae* family, of which hakeas are a member. Rather, it was a Western woody pear, or *Xylomelum occidentale*. The name stems from the Greek words *xylon*, meaning "wood," and *melon*, meaning "apple" or fruit more generally. *Occidentale*, meanwhile, means "Western."

Rather than proving detrimental, Molloy's lack of knowledge about plant identification meant that she was receptive to the Noongar name for this plant, *danja*. When she was collecting, Molloy used anyone who would accompany her, including Noongar guides. She wrote to Mangles that "[t]he Soldiers who used to pass between this, and Augusta unmolested and unencumbered with any thing but their knapsacks, are now seen to bring from thence Specimens of all sorts of Plants under their Arms. The Native Herdsmen are also employed bringing in some desired 'Plant' or 'Fruit.'"[69] Her awareness of the name *danja* indicates that she listened to her Noongar guides. Indigenous Australian plant collectors are "often portrayed as silent partners" in published accounts from the colonial period,[70] but through Molloy's letter they still speak to readers, reminding them that, as Mahood found, there is "some other kind of framework for knowing this country."[71]

This mode of knowing was developed after the first modern humans arrived in Australia approximately 60,000 years ago towards the end of the Pleistocene period, when glaciation meant that sea levels were lower and land bridges allowed them to travel onto the continent from Asia. Over thousands of years, they refined their knowledge about their country to sustain themselves physically and psychically. Blossoms on gum trees or the movement of stars signaled coming seasons and weather patterns;[72] plants were used extensively for medicine; and specific sites were honored with cultural ceremonies to acknowledge the land's role in providing sustenance. Noongar people were taught to collect only what they needed at a particular time; for example, their law "required that no seed-bearing plants be dug up after flowering."[73] They used infusions of *Xylomelum occidentale*'s leaves and bark to relieve pain, and roasted and ate seeds from the fruit.[74]

Yet *Xylomelum occidentale* also has a history that precedes humans. *Xylomelum*'s family is *Proteaceae*, which, over 115 million years, has branched into over eighty genera and seventeen hundred species.[75] These species occur predominantly in the southern hemisphere, particularly southern Africa and the SWAFR. Its diversification has been prompted by "a long history of harsh environment, recent disturbance and a plethora of interactions with fauna,"[76] as well as extensive interaction with fire.[77]

By considering *Xylomelum occidentale* in terms of its long lineage and its relationship with Indigenous and non-Indigenous Australians, one gathers an appreciation for this remarkable plant. Its life story shakes us from a perception of the plant purely in terms of its aesthetics and utility, and compels us to consider it as an agent in its own right. It has drawn humans and other-than-humans into its orbit, whether for physical, emotional, or aesthetic nourishment. It has drawn me, too, into its story, to illuminate its importance for Noongar peoples and for lovers of flowers such as Molloy. Although the human time in its narrative is but a pinprick, to expand on its interactions with humans is to show the plant's influence and agency. *Xylomelum occidentale* thus encourages us to ask, as Edward O. Wilson does in *Biophilia*, his iconic exploration of humans' affinity for the natural world, "Is it possible that humanity will love life enough to save it?"[78]

This question is critical in terms of the survival of the SWAFR in the Anthropocene. The area remains under threat from land clearing (which has already "given rise to the largest single extirpation event in Australia"[79]) for agriculture and urbanization, and the introduction of *Phytophthora cinnamomi*, a soil-borne mold that produces root rot or dieback. These are the main causes of declines in plant diversity in this area, but other threats include weeds, grazing, the prohibition of traditional Indigenous burning practices,[80] and the loss of local knowledge through the dispossession of Noongar peoples from massacre and disease. Climate change has also been recognized as a threat, with the region predicted to become hotter and drier.

Perhaps it is time to recognize that our selves are not our own; they are intimately bound up with and constituted by millions of other lives. The dwarf hammer orchid cannot exist without the thynnine wasp, just as Molloy could not have derived emotional and intellectual nourishment from the verticordias, kingias, nuytsias, and pimelias that she collected. Amid the countless deaths of living organisms that mark the Anthropocene, it is critical that we write, recognize, and critique our awareness of the ways in which human selves extend into ecosystems, and how our ecosystems stretch into, shape, and support a myriad of human and other-than-human lives.

Funding

This research was supported by an Australian Research Council Discovery Early Career Researcher Award [DE160100238].

The University of Queensland

Disclosure Statement

No potential conflict of interest was reported by the author.

Notes

1. Hopper and Gioia, "Southwest Australian Floristic Region," 628.
2. *Oxford English Dictionary.*
3. Swarts and Dixon, "Terrestrial Orchid Conservation," 545.
4. Myers et al., "Biodiversity Hotspots," 853.
5. See Bennett, *Vibrant Matter*; Haraway, *Staying with the Trouble*; and Tsing, *The Mushroom.*
6. Lists of Molloy's specimens are written into James Mangles' letters books. See Molloy, Letters.
7. It is interesting to note that another Australian female scientist, Edith Coleman (1874–1951), confirmed and built upon international research on the pollination of orchids through copulation. See Clode, *The Wasp.*
8. Smith and Watson, *Reading Autobiography* (2nd ed.), 1.
9. Anderson, *Autobiography*, 4.
10. Whitlock, "Post-ing Lives," vi.
11. Braidotti, *The Posthuman*, 1.
12. Lines, *An All Consuming Passion.*
13. Molloy, Letter to Frances Birkett.
14. Molloy, Letter to Elizabeth Kennedy.
15. Molloy, Letter to James Mangles, September 1838.
16. Shteir, *Cultivating Women*, 36.
17. Molloy, Letter to James Mangles, March 21, 1837.
18. Molloy, Letter to James Mangles, January 25, 1838.
19. Molloy, Letter to Elizabeth Besley.
20. Molloy, Letter to James Mangles, June 30, 1840.
21. Molloy, Letter to James Mangles, June 4, 1841.
22. Hasluck, *Portrait with Background.*
23. Molloy, Letter to James Mangles, June 4, 1841.

24. Molloy, Letter to James Mangles, January 28, 1838.
25. Molloy, Letter to James Mangles, January 31, 1840.
26. Konchar Farr and Snyder, "From Walden Pond," 198.
27. Konchar Farr and Snyder, "From Walden Pond," 203.
28. Smith and Watson, *Reading Autobiography*, 268.
29. In 1998, Konchar Farr followed her essay with "American Ecobiography," an analysis of Edward Abbey's *Desert Solitaire* and Terry Tempest Williams' *Refuge*. Describing ecobiographies as residing in "the space in which American autobiography and nature writing meet," she relates how the desert melds with Abbey's solitude and construction of self, and how Tempest Williams' grief on her mother's death leads to an imagined fusing with her mother and a maternal motivation to protect the earth for future generations (94). Perreten's 2003 essay "Eco-Autobiography: Portrait of Place|Self-Portrait" examines how the structure and texture of three works are shaped by the author's interaction with a setting. In 2010, Edlich published "Richard K. Nelson's *The Island Within*: Environmental Life Writing as Ecological Identity Work," an essay that explores how Nelson's text captures the shift from "an anthropocentric conception of the Enlightenment Self and its traditional attendant modes of autobiographical self-representation to a highly individualised ecocentric understanding of human identity" (204). In 2011, Edlich expanded his analysis of ecobiographical texts by contemplating works by ecocritics Ian Marshall and John Elder, which are "complex composite texts that transgress the boundaries of autobiographical environmental literary scholarship and full-fledged 'ecobiographical' or 'eco-autobiographical' representation" ("Autobiographical Ecocritical Practices," 930). In 2011, Straight published *Autobiography, Ecology and the Well-Placed Self* and referenced a resurgence of regionalists in the US who are producing personal and local stories in response to "our modern condition of environmental and social alienation" (11). These "writers-in-place," he continues, "candidly examine their intimate connections to regional mythologies and recognise the inseparability of personal narratives from the large histories of region and nation" (12). Straight labels this mode "natural biography" and includes a bibliography of some fifty texts as exemplars. Kennedy's 2012 essay "Humanity's Footprint: Reading 'Rings of Saturn' and 'Palestinian Walks' in an Anthropocene Era" in the "posthuman lives" issue of *Biography* also contemplates the indivisibility of the self with a specific region. Drawing on biogeography, or the way in which geography shapes life forms, Kennedy describes Sebald's *The Rings of Saturn* as "infused by a sensibility that sees the natural and the human as deeply intertwined" (176). This essay's focus on climate change is a welcome—and ultimately inevitable—step in analyses of ecobiography. The most recent mention of ecobiography is in Hornung's foreword to a collection of essays, *Ecology and Life Writing* (ix). While the form is not dwelled upon at length, the volume's wide range of Eastern, Western, and Pacific perspectives on the literary representation of environment, and how environment shapes textual selves, points to the accelerating transcultural interest in this field.
30. In Indigenous Australian English, "country," as Rose explains in *Nourishing Terrains*, is "not only a common noun but also a proper noun" (7). Country

is not used in the sense of "a day in the country"; rather, it is "a living entity" (7).

31. Rose, *Nourishing Terrains*, 10.
32. It is important to note that Rose's descriptions of Indigenous peoples' relationships to country do not apply to all Indigenous Australian peoples. The Australian Bureau of Statistics indicates that the majority (eighty-one percent) of Indigenous Australian people currently live in cities and non-remote areas. However, these relationships were a fundamental tenet for Indigenous Australian peoples at the point of the British invasion and remain so for many Indigenous Australian peoples today.
33. Rose, "Val Plumwood's Philosophical Animism," 95.
34. Roughsey, *Moon and Rainbow*, 42–43.
35. Grossman, *Entangled Subjects*, 63.
36. These interactions have been described by Bunurong, Yuin, and Palawa man Bruce Pascoe in his seminal *Dark Emu, Black Seeds*, and in Gammage's *The Biggest Estate on Earth*.
37. The southwest Western Australia area contains several linguistic subgroups. For example, Whadjuk Noongars are the traditional custodians of the area surrounding what is now known as Perth; Wardandi Noongars are the traditional custodians of what is now the area between Cape Leeuwin and Cape Naturaliste; and Wirlomin Noongars are the traditional custodians of the area around what is now known as Albany. The Noongar people are one of many regional groups, or "nations," of Indigenous people in Australia.
38. John Molloy was also heavily involved in a massacre carried out against Wardandi Noongars. See White, "'Paper Talk.'"
39. These ecobiographies tend to be concerned with the impact of European land-management methods on a non-European environment with vastly different soil, rainfall, plants, and animals. See, for example, Mahood, *Craft*; Muecke, *No Road*; Somerville, *Body/Landscape Journals*; Beudel, *A Country in Mind*; and Weldon, *The Lake's Apprentice*. Critical commentary on these ecobiographies has been published in Brennan, "Kim Mahood's *Craft*"; Probyn, "Poetics of Failure"; Probyn and Somerville, "'Postcolonial Practice of Writing'"; Slater, "Intimate Australia" and "Is Any Body Home?"; and White, "'I Felt This Landscape." These ecobiographies differ from the numerous accounts of humans observing, enjoying, and deriving respite from the natural world, often in the form of grief or rambler narratives (such as *H Is for Hawk* by Macdonald or Strayed's *Wild*), in that they highlight the connections between humans and vital ecosystems. They are also different to natural histories of a place, which likewise do not examine the way in which human selves are shaped by other-than-human selves.
40. Simpson, "Writing the Country," 273.
41. Mahood, *Position Doubtful*, 2.
42. Mahood, *Position Doubtful*, 3.
43. Mahood, *Position Doubtful*, 2.
44. Mahood, *Position Doubtful*, 5.
45. Mahood, *Position Doubtful*, 1.
46. Mahood, *Position Doubtful*, 5.
47. Mahood, *Position Doubtful*, 5.
48. Mahood, *Position Doubtful*, 29.
49. Mahood, *Position Doubtful*, 28.

50. Mahood, *Position Doubtful*, 213.
51. Mahood, *Position Doubtful*, 213.
52. Mahood, *Position Doubtful*, 213.
53. Mahood, *Position Doubtful*, 214.
54. Mahood, *Position Doubtful*, 214.
55. Mahood, *Position Doubtful*, 180.
56. Mahood, *Position Doubtful*, 188.
57. Mahood, *Position Doubtful*, 193. In Indigenous Australian culture, a songline, also known as a dreaming track, is a path across the country or sky created during the Dreamtime. These paths are recorded in traditional songs, stories, dances, and artwork.
58. Canning Stock Route Art Project, "Billy (Mr. P) Patch."
59. Mahood, *Position Doubtful*, 214.
60. Mahood, *Position Doubtful*, 33.
61. Mahood, *Position Doubtful*, 7.
62. Mahood, *Position Doubtful*, 162.
63. Mahood, *Position Doubtful*, 2.
64. Molloy, Letter to James Mangles, January 25, 1838.
65. Molloy, Letter to James Mangles, January 25, 1838.
66. Molloy, Letter to James Mangles, 1840.
67. Molloy, Letter to James Mangles, March 14, 1840.
68. Molloy, Letter to James Mangles, March 14, 1840.
69. Molloy, Letter to James Mangles, January 31, 1840.
70. Clarke, *Aboriginal Plant Collectors*, 7.
71. Mahood, *Position Doubtful*, 162.
72. Hansen and Horsfall, *Noongar Bush Medicine*, x, 3.
73. Hansen and Horsfall, *Noongar Bush Medicine*, xi, 3.
74. Hansen and Horsfall, *Noongar Bush Medicine*, 216.
75. *Xylomelum* is a genus which consists of seven species, of which *Xylomelum occidentale* is one. A genus is ranked above the species but below the family.
76. Pausas and Lamont, "Ecology and Biogeography," 1472–1473.
77. Pausas and Lamont, "Ecology and Biogeography," 1474–1475.
78. Wilson, *Biophilia*, 145.
79. Swarts and Dixon, "Terrestrial Orchid Conservation," 545.
80. Monks et al., "Threatened Plant Species," 60.

ORCID

Jessica White http://orcid.org/0000-0002-9291-5712

Works Cited

Anderson, Linda R. *Autobiography*. Abingdon: Routledge, 2011.

Australian Bureau of Statistics. "3238.0.55.001—Estimates of Aboriginal and Torres Strait Islander Australians, June 2016." https://www.abs.gov.au/ausstats/abs@.nsf/mf/3238.0.55.001

Barry, Bernice. *Georgiana Molloy: The Mind That Shines*. Witchcliffe, WA: Redgate Consultants, 2015.

Bennett, Jane. *Vibrant Matter: A Political Ecology of Things*. Durham, NC: Duke UP, 2010.
Beudel, Saskia. *A Country in Mind: Memoir with Landscape*. Crawley, WA: U of Western Australia P, 2013.
Braidotti, Rosi. *The Posthuman*. Cambridge: Polity, 2013.
Brennan, Bernadette. "Kim Mahood's *Craft for a Dry Lake*: A Work in Progress." *Southerly* 66, no. 1 (2008): 91–105.
Canning Stock Route Art Project. "Billy (Mr. P) Patch." Accessed May 1, 2019. http://www.canningstockrouteproject.com/artists-and-artworks/billy-mr-p-patch/
Clarke, Philip A. *Aboriginal Plant Collectors: Botanists and Australian Aboriginal People in the Nineteenth Century*. Dural, NSW: Rosenberg, 2008.
Clode, Danielle. *The Wasp and the Orchid: The Remarkable Life of Australian Naturalist Edith Coleman*. Sydney, NSW: Pan Macmillan, 2000.
Edlich, Micha. "Autobiographical Ecocritical Practices and Academic Environmental Life Writing: John Elder, Ian Marshall, and Catriona Mortimer-Sandilands." *Literature Compass* 8, no. 12 (2011): 929–940.
Edlich, Micha. "Richard K. Nelson's *The Island Within*: Environmental Life Writing as Ecological Identity Work." *Auto/Biography Studies* 25, no. 2 (2010): 203–218.
Gammage, Bill. *The Biggest Estate on Earth: How Aborigines Made Australia*. Crows Nest, NSW: Allen and Unwin, 2011.
Grossman, Michèle. *Entangled Subjects: Indigenous/Australian Cross-Cultures of Talk, Text, and Modernity*. Amsterdam: Rodopi, 2013.
Hansen, Vivienne, and John Horsfall. *Noongar Bush Medicine: Medicinal Plants of the South-West of Western Australia*. Crawley, NSW: U of Western Australia P, 2016.
Haraway, Donna. *Staying with the Trouble: Anthropocene, Capitalocene, Chthulucene*. Durham, NC: Duke UP, 2016.
Hasluck, Alexandra. *Portrait with Background*. Melbourne, VIC: Oxford UP, 1955.
Hopper, Stephen D., and Paul Gioia. "The Southwest Australian Floristic Region: Evolution and Conservation of a Global Hot Spot of Biodiversity." *Annual Review of Ecology, Evolution and Systematics* 35, no. 1 (2004): 623–650.
Hornung, Alfred, and Zhao Baisheng, ed. *Ecology and Life Writing*. Heidelberg: Winter, 2013.
Kennedy, Rosanne. "Humanity's Footprint: Reading 'Rings of Saturn' and 'Palestinian Walks' in an Anthropocene Era." *Biography* 35, no. 1 (2012): 170–189.
Konchar Farr, Cecilia. "American Ecobiography." In *Literature of Nature: An International Sourcebook*, edited by Patrick D. Murphy, Terry Gifford, and Katsunori Yamazato, 94–97. Chicago, IL: Fitzroy Dearborn, 1998.
Konchar Farr, Cecilia, and Philip A. Snyder. "From Walden Pond to the Great Salt Lake: Ecobiography and Engendered Species Acts in *Walden* and *Refuge*." In *Tending the Garden: Essays on Mormon Literature*, edited by Eugene England and Lavina Fielding Anderson, 197–212. Salt Lake City, UT: Signature, 1996.
Lines, William J. *An All Consuming Passion: Origins, Modernity, and the Australian Life of Georgiana Molloy*. Berkeley, CA: U of California P, 1994.
Macdonald, Helen. *H is for Hawk*. New York, NY: Grove Press, 2014.
Mahood, Kim. *Craft for a Dry Lake*. Sydney, NSW: Anchor, 2000.

—— *Position Doubtful: Mapping Landscapes and Memories*. Brunswick, VIC: Scribe, 2016.

Molloy, Georgiana. Letter to Elizabeth Besley, November 7, 1832. *ACC 3278A. Papers of Georgiana Molloy. J. S. Battye Library of West Australian History*, Perth.

—— Letter to Elizabeth Kennedy, c. November 1830. DKEN3/28/9. Kennedy Family of Crosby House, Crosby-on-Eden. Carlisle Archive Centre.

—— Letter to Frances Birkett, April 15, 1831. DKEN3/28/9. Kennedy Family of Crosby House, Crosby-on-Eden. Carlisle Archive Centre.

—— *Letters to James Mangles. MN879. Papers of Captain James Mangles. J. S. Battye Library of West Australian History*, Perth.

Monks, Leonie, Sarah Barrett, Brett Beecham, Margaret Byrne, Alanna Chant, David Coates, J. Anne Cochrane, Andrew Crawford, Rebecca Dillon, and Colin Yates. "Recovery of Threatened Plant Species and Their Habitats in the Biodiversity Hotspot of the Southwest Australian Floristic Region." *Plant Diversity* 41, no. 2 (2019): 59–74.

Muecke, Stephen. *No Road (Bitumen all the Way)*. Fremantle, WA: Fremantle Arts Centre, 1997.

Myers, Norman, Russell A. Mittermeier, Cristina G. Mittermeier, Gustavo A. B. da Fonseca, and Jennifer Kent. "Biodiversity Hotspots for Conservation Priorities." *Nature* 403, no. 6772 (2000): 853–858.

Oodgeroo Noonuccal, and Bronwyn Bancroft. *Stradbroke Dreamtime*. Sydney, NSW: Angus and Robertson, 1993.

Oxford English Dictionary. Oxford: Oxford UP, 2000.

Pascoe, Bruce. *Dark Emu, Black Seeds: Agriculture or Accident?* Broome, WA: Magabala, 2014.

Pausas, Juli G., and Byron B. Lamont. "Ecology and Biogeography in 3D: The Case of the Australian Proteaceae." *Journal of Biogeography* 45 (2018): 1469–1477.

Perreten, Peter F. "Eco-Autobiography: Portrait of Place|Self-Portrait." *a/b: Auto/Biography Studies* 18, no. 1 (2003): 1–22.

Probyn, Fiona. "A Poetics of Failure Is No Bad Thing: Stephen Muecke and Margaret Somerville's White Writing." *Journal of Australian Studies* 26, no. 75 (2002): 17–26.

Probyn, Fiona, and Margaret Somerville. "Towards 'A Postcolonial Practice of Writing.'" *Hecate* 30, no. 1 (2004): 56–71.

Rose, Deborah Bird. *Nourishing Terrains: Australian Indigenous Views of Landscape and Wilderness*. Canberra, ACT: Australian Heritage Commission, 1996.

—— "Val Plumwood's Philosophical Animism: Attentive Interactions in the Sentient World." *Environmental Humanities* 3, no. 1 (2013): 93–109.

Roughsey, Dick. *Moon and Rainbow: The Autobiography of an Aboriginal*. Sydney, NSW: Reed, 1971.

Scott, Kim, and Hazel Brown. *Kayang and Me*. Fremantle, WA: Fremantle Arts, 2005.

Shteir, Ann B. *Cultivating Women, Cultivating Science: Flora's Daughters and Botany in England, 1760–1860*. Baltimore, MD: Johns Hopkins UP, 1996.

Simpson, Inga. "Writing the Country." *Griffith Review* 63 (2019): 272–281.

Slater, Lisa. "Intimate Australia: *Body/Landscape Journals* and the Paradox of Belonging." *Cultural Studies Review* 13, no. 1 (2011). http://epress.lib.uts.edu.au/journals/index.php/csrj/article/view/2160

—— "Is Any Body Home? Rewriting the Crisis of Belonging in Margaret Somerville's *Body/Landscape Journals*." *Southerly* 62, no. 1 (2002): 181–192.

Smith, Sidonie, and Julia Watson. *Reading Autobiography: A Guide for Interpreting Life Narratives*. Minneapolis, MN: U of Minnesota P, 2001.

—— *Reading Autobiography: A Guide for Interpreting Life Narratives*. 2nd ed. Minneapolis, MN: U of Minnesota P, 2010.

Somerville, Margaret. *Body/Landscape Journals*. North Melbourne, VIC: Spinifex, 1999.

Straight, Nathan. *Autobiography, Ecology, and the Well-Placed Self: The Growth of Natural Biography in Contemporary American Life Writing*. New York, NY: Peter Lang, 2011.

Strayed, Cheryl. *Wild: A Journey from Lost to Found on the Pacific Crest Trail*. New York, NY: Alfred A. Knopf, 2012.

Swarts, Nigel D., and Kingsley W. Dixon. "Terrestrial Orchid Conservation in the Age of Extinction." *Annals of Botany* 104, no. 3 (2009): 543–556.

Tsing, Anna. *The Mushroom at the End of the World*. Princeton, NJ: Princeton UP, 2015.

Weldon, Annamaria. *The Lake's Apprentice*. Crawley, NSW: U of Western Australia P, 2014.

White, Jessica. "'I Felt This Landscape Knew I Was There': *The Lake's Apprentice* and Ecobiography." In *Offshoot: Contemporary Life Writing Methodologies and Practice*, edited by Donna Lee Brien and Quinn Eades, 121–135. Crawley, NSW: U of Western Australia P, 2018.

—— "'Paper Talk': Testimony and Forgetting in South-West Western Australia." *JASAL: Journal of the Association for the Study of Australian Literature* 17, no. 1 (2017): 1–13.

Whitlock, Gillian. "Post-ing Lives." *Biography* 35, no. 1 (2012): v–xvi.

Wilson, Edward O. *Biophilia*. Cambridge, MA: Harvard UP, 1984.

Period Rhetoric, Countersignature, and the Australian Novel

By Thomas Bristow

ABSTRACT
This article explores literary practices placing the writer in dialogue with the places he has inhabited recently while researching the Australian novel. This includes a fictocritical engagement with place-based Australian literature (via Xavier Herbert and Randolph Stow) and a maverick whizz through deconstruction and genre studies. Written in an elegiac mode punctuated by an environmental humanities countersignature, this example of period rhetoric embodies autobiography in the Anthropocene, the event horizon of human signature.

"It does not seem to me," declares Austerlitz, "that we understand the laws governing the return of the past, but I feel, more and more as if time did not exist at all, only various spaces interlocking according to the rules of a higher form of stereometry, between which the living and the dead can move back and forth as they like."

— W. G. Sebald, *Austerlitz*

The pioneers had difficulty in establishing permanent settlements, having several times to abandon ground they had won with slaughter and go slaughtering again to secure more. This abandoning of ground was due not to the hostility of the natives, hostile enough though they were, but to the violence of the climate, which was not to be withstood even by me so well equipped with lethal weapons and belief in the decency of their purpose as Anglo-Saxon builders of empire.

— Xavier Herbert, *Capricornia*

Prelude

I do my half best not to disturb the sleeping kelpie as I get up and switch off the fan. The extra degree of peace from that small event this late evening in a single hotel room is welcome. It opens me up to the music from stereos in cars gliding past on Flinders Street and to the last of the night's rain rushing down the roof of the empty Sovereign Hotel under yet another majestic star-filled night. With the fan going, you can mistake the sound of the wind fingering the rusting advertising boards of the car lot opposite for that of birds in the palm trees. Twenty-three is my lucky number and I left the novel I was reading open at that page so I could return to the lines in the middle of the page when I woke: "Rain stretched down like silver wires from heaven of pitch to earth of seething mud."[1] Before I knew about the hydrologic cycle or climate emergency, as a boy I used to hold out a tape recorder from my bedroom window and catch the sounds of rain to replay when it was over. That house was one street away from where four generations of my mother's family were born. The village green is soaked in Cooper cricket history; the cemetery is nearby. I captured some rain from my mother's house (our last house in England) before I left the East Midlands rural belt for the final time last year. But rain means something quite different in the dry tropics as an adult. The end of rain has been on my mind the last few days as we are changing to the dry season in North Queensland. And rain comes to me again as I look at the fan's blades spinning with the relinquished energy of non-renewables; the temporary, hyperlocal termination of the coal-fired electricity supply translating ancient phytoplankton at my wall seems to be poignant for the blades' diminishing rotation is subject only to their presence in the gravity field.

It's just past 2 a.m. and the rain has come again. I wake with the desire to see the road outside soaking, perhaps for the last time for some months. I think that this third day of rain, the most significant spell in the last two months, will remind residents of the January flood that placed North Queensland on national news and on media channels around the world. And that devastating event, to the white elders that I've met here, is a reminder of a handful of other drastic weather events in Townsville in the last fifty years. The signatures of earth's screaming are captured in oral history and passed on through family and community in the depleting resilience of the subtropics of Australia. Partly desensitizing folk to the losses to come, partly reminding them of drastic climate change in living memory, such events and their responses are a common fixture.[2]

Preface

While considering the future and endpoint of autobiography, Paul John Eakin reflects on its emergence while new concepts of time and person provided a rationale for its form and content.[3] Looking to the discipline's future, Eakin considers the rise of social media and its impacts on "social narration" in the digital age, but there is no mention of environmental despoliation, climate crisis, or species extinction in his imagining. This critical blindness, however, comes with insight on new forms providing "solace for our desire to work out our place in the larger scheme of things."[4] The sole qualification to this positivity, I feel, would be to say that narrative coupling of time and identity under the sign of the Anthropocene would do more than permit us to pay attention to others: it adds urgency to our writing practices, spelling out the inseparability of human history and environmental history, refiguring our identity in geological time, and foregrounding impoverished more-than-human subjects and despoiled sites of subjectivity, of which we have a remarkable literary record.

Jacques Derrida's attack on the Kantian tradition's project to build a bridge between the supersensible and its signs is something I wish to imitate in this article, with pictures invoking Sebaldian stereometry in mind.[5] Sebald's use of narrative and image invokes a site of affirmation and betrayal wherein the apparent presence of being (and its origin) is replaced in a situation of temporal, historical, and ontological disjunction.[6] Derrida sees the event of hanging a picture in a gallery as an act of countersignature, writing against the world that exists (betrayal) in its promotion of what comes before (affirmation); the act of a viewer accessing the framed work of art is understood as countersignature, too. For David Wills, countersigning "means writing one's name as it were on the work, even if by proxy, performing some sort of autobiographical, and thus testamentary or obituary function."[7] We might ask how this obituary function will present itself as a manifestation of social interaction and activity in Anthropocene life writing, rather than originating in some transcendent rhetorical category, "already there" in the world (such as elegy, for example).

In order to answer this, my exploration of countersignature follows Roland Barthes' analysis of the press photograph (1961), seeking a dialogue that endeavors to identify the contours of contiguity between analogical content (dog, portrait, landscape) and connoted message (epic loss).[8] Unlike the analgon, whose denoted message is indebted to its style of representation, the secondary meaning of my photographs is indebted to the way its "signifier is a certain 'treatment' of the image."[9] Anthropocene autobiography is the period rhetoric that determines the connotation procedure in this article, not only coding the photographic analogue but opening up the structural autonomy of its "photogenia" to the technical effects of juxtaposition (past and present images) and allusion (to embodied experience and reading the novel). This approach brings spontaneity, imprecision, and fleetingness to our voices; as Paoloa

Bilborough notes, "non-literal, poetic approaches to textual composition (such as fictocriticism) can ethically frame life stories by incorporating shared privacies, obscure emotional insights, and sensitive cultural analysis."[10] The nonliteral form of stereometry that I employ is of an elegiac mode alert to ethical framing, owing to my interest in loss and my interest in the literary record that speaks not to change but to contiguity. Such a record marks the end of the relevance of the tools of history. Deborah Bird Rose argues for the necessity of bringing together the world before us and the world that is to come, so that they are not understood as contrasts but more openly stand in a dialectical relationship to each other. Rose tells us that we need to drop narratives of the privatized transforming self and feel okay about locating our "selves" within a world enlivened by mutualism that entails attentiveness "to the lives of a myriad selfhoods" and "intersubjective dialogues."[11] I come to a sense of an inconsistent and numbered self through the impact of transportation in physical space and through finding comfort in Sebald's interlocking spaces. I make a single demand on the reader: to allow these images and descriptions to sit loosely in the diaristic form in the hope they will resonate with each other during slow reading that holds the significance of image and text open. Let's allow for porosity in things, even writing.

Anthropocene Field Notes 1

Alert to writerly sensitivity, Heather Kerr has noted that fictocriticism models a practice of empathy; its tone and form "belong to the larger category of 'trauma studies', including the feelings left out of historical narratives."[12] I feel this keenly and turn to others to navigate the space of Anthropocene anxiety in words. Rose's mentoring has helped me travel to the eye of the storm, for she (with other Australians here) has commenced the great howling. In a move from intellectualizing (and thus extending) substitute grief to being overwhelmed by environmental holocaust on the journey to a 4 °C rise in global temperature, Rose locates a new source for Anthropocene autobiography: "the howling starts within empathy, grief, and much more, and it reverberates beyond us."[13] I take note.

The Anthropocene is in all our lives, inflected in specific ways owing to our particular personal experiences, themselves forged on the multispecies lintels of ecology and separating out in distinct affects. I am thinking of multiple public affects from one community spilling into pools that separate out like oil and water while my dog Marx instinctively follows the scent lines of the pavement and decodes the messages of hundreds of animals while recalling the precise geolocation and time stamp of their last twenty piss patches. I march single-purposely to the laundromat.

Figure 1. Marx at the Eddie Kaiko Mabo Library, 2019. Image copyright the author.

During a quick tour of James Cook University, I wanted to see what he would make of the grounds of the Eddie Kaiko Mabo Library, where he's strictly not allowed. The building's style, largely influenced by the architectonics of Captain James Cook's *Endeavour*, has been countersigned through a renaming ceremony inscribing the building with the decolonizing vision of the cultural and political achievements of a former groundsman and gardener here. This seemed to mean little or nothing to Marx in his desire to rush through the green space and disturb the stone bush curlews (Figure 1).

Joshua Lobb marks the literary flipside of the long-standing philosophical concern for the impossibility of inhabiting another species' world: "would I always be ruled by my own anthropocentrism?"[14] For a moment, I wonder whether I'm doing justice to Marx's subjectivity in my anthropocentric reading of his canine-centrism. I should be writing towards "interactions that permit the location and interests" of both humans and animals.[15] Equally, should I not ask: As a dog, is it possible for him to inhabit the world of a researcher? He certainly knows what I'm feeling when I jump from the writing desk in accomplished delight. He reads significance in my body returning from the library or shifting from the workspace to the bathroom. There's insignificance for him as I edit a footnote for Chicago Style as his design is to go outside, quite opposed to mine, which is staying put and finishing the piece (although there's some middle ground that should be parenthesized here: right now, he's sitting on my foot, complicit in the work and waiting for change). Perhaps I'm stubborn, but he isn't deterritorializing my subjectivity; that's something for theorists to ruminate over. He'll just get me out of this chair on the hotel's front verandah a little quicker than I would on my own. It's not a combative relationship (at this very moment of typing); he's keeping me in shape. In his own way of navigating the day, he'll get up and dance to let me know it's right to go. Why refuse when he has such good timing?

I find myself collocating the single life I'm legally responsible for (other than my own) with the sense of responsibility we have for all other lives (including flora and fauna) that this place constantly reminds me of. A cluster of two inner-city suburbs of Townsville is at present a small spot in which I reside in a closed hotel with those turned homeless by the summer flood event, by alcoholism, and by mental illness. It's where I walk Marx. As the images show, the building is close to a long-standing space of European faith and community. It's telling, I know, but I hope our continents can join once more in peace.

The event of the writing you read today took place while contrite Byzantine a cappella chanting filled the interior space of the postwar, Romanesque-revival-styled church, Saints Theodores, Townsville, this late summer (Figures 2 and 3). We might turn to the twentieth-century Australian novel (Randolph Stow, for example) if we wish to think more on the subtleties of old civilizations and their impact on Australia's (fast)

Figure 2. Saints Theodores Church, Townsville, 2019. Image copyright the author.

Figure 3. Flinders Street, Townsville, 2019. Image copyright the author.

Figure 4. Damage to the Mansfield Hotel by Cyclone Althea, 1971. Flickr, Townsville. CityLibraries. CC BY-SA 2.0 https://www.flickr.com/photos/citylibrariestownsville/9519474067/

development as it is caught up in a specific region. Behind the church behind the hotel the sun revolves Cootharinga; a pink granite monolith, just meters short of mountain classification, which is home to the area's radio communication installations. Molten lava solidified deep underground to create the most distinct natural feature in Townsville, itself marked as a monument to Captain Robert Towns, the disputed founder of the once-named settlement Castle Hill. Curiously, Balmain (an inner-west suburb of Sydney) is where Towns was buried, and where I last laid my suitcase and hat on the way from New South Wales to Queensland.

Figure 5. Flooded Streets, South Townsville. Flickr, Rob and Stephanie Levy. CC BY 2.0 https://www.flickr.com/photos/robandstephanielevy/5413071408/

Towns' plaque sits in a cleared space at the top of the hill a short distance from an ammunition magazine, on the slopes of kangaroo grass, among vine thickets, privately planted banyans from the 1890s, and fifty species of introduced flora that were ravaged, along with the natural vegetation, by goats until the late 1930s. It's one of those exceptional Australian knots that, once unraveled, would expose a master mariner's involvement in open-sea (pelagic) whaling trades, pastoralism, cotton, and the exploitation of imported Chinese and Indian laborers on plantations. It's easy to get caught up in this.

Countersignature 1

Only a Queenslander miner's cottage. Only an Australian-built, 1961 Holden Special EK (Figure 4). Only the whitening of the sky from a transformative cyclone over a dwelling place whose security is mortgaged to any future reconfiguration of relations between the petrochemical industry and climate emergency. Queensland architectural prettiness is unmistakable—a nostalgic metonym for Australian postwar confidence (despite the rise of nuclear) linked to the safety and security felt in some regional working-class areas before the new fear around jobs and climate came into view (Figure 5). In North Queensland, this has recently bifurcated into an enduring oppositional rhetoric of coal versus jobs that equally dismiss non-human species. Incidentally, that countersignature to the bending trees was edited immediately after my first experience of *tidal friction*—that is, my walking the moon-drenched, flooded space of Ross Creek before returning to the closed hotel that sits at the center of ten car lots in between streets either side of the building named after two English explorers (Charles Sturt and Matthew Flinders).

I bring together these images and references to colonial legacies of our resource-extraction diet on the eve of a federal election as local issues are at the center of national debate now—the specific impact of climate change here in this region and the establishment of new coalmines and their impact on an at-risk species (the black-throated finch) is one case in point.

I'm keen to know more about this finch, but I'm finding it hard to meet myself in the present. And I need just that right now (meeting myself) on the back of Brexit, the loss of parents, and the transequatorial relocation this January 1. I turn to Stow's *The Merry Go Round in the Sea* for comfort and inspiration. The young protagonist, Rob, is experiencing a world moving away from the naïve equilibrium he wishes to sustain for eternity. As Rob's relationship with the Western Australian beach changes in this awakening to the Japanese threat to the nation, Stow crafts short, rich, epigrammatic observations of the diurnal passage in a somber register: "the sunlight changed. Something came over the sun, a kind of dusk."[16] And yet, the text seems to resist this fall for it is full of the present moment before the dusk; as vulnerable and temporary as it might feel, it is captured in words and shot into carbon for our enjoyment. I find this specific open dialogue and its contagious affective temporality.

In Townsville, as I attune to Rob's emotional latitude of losing, nonhuman agency swings into view: palms, eucalypts, the church, the sky. Mangos and wild shrubs scrape possessively against the saltwater creek I sense behind the railway yards by the hotel. To the north and south, the beaches move in the wind. Each individual cool night gnaws a little from the tropical humidity. Time is irredeemable. And far to the north is Capricornia. The colonially constructed outback is accessible to the west in under two hours. The Great Barrier Reef is much closer; you can smell it through the stinking passionfruit and the sickening strangulated frangipani. I feel calm out walking by the river, despite the toxic politics in the city. At the very least, the seeds of memories to be enjoyed in later life are rushing through swaying weeping figs amongst royal, cabbage, and Alexandra palms that follow the creek into the river a long way out of town. They reach towards veranda roofs and their scraping sound can be heard after the birds have wheeled through the late sun and settled for the evening. I turn away from the crocodile sign at a short crop of eucalypts and a worker's cottage appears in the starlit rectangle shaped by the trees. Yes, it's a nineteenth-century miner's cottage. Coconut pods the size of your head lie at these European feet resting among the palm detritus from a year's worth of rain in a January week when Australia recorded its hottest month since records began. It's dizzying. I'm looking for the sea almond but I come across flowers of a color I have no name for; I mistake fruits for citrus, green on stalks of bright orange and fail to recognize the passionfruit vine. Since the floods, it's a very green town, with streets and houses thick with dark-green milky succulents under the lush, smooth, oversized leaves. You feel like they and you might stay here always, and it will be today forever.

Anthropocene Field Notes 2

This particular sector just south of the city boundary was carved out by the unstoppable flows of water and immovable motivations of town planners in the last century. Shortly before waking to the rain in the middle of the night, I read of a boat in Xavier Herbert's great novel *Capricornia*, drifting before "a dying breeze into the creek up which she had stolen with Civilization years before."[17] I imagined the scene as I drifted to sleep, and it came back to me as I switched off the fan to help me connect to Flinders Street a short walk from here. Once a dumping site for the tropical town following the damage caused by Cyclone Alethea in December 1971, this slice of Townsville is defined by a gorgeous stretch of water that empties into the Coral Sea, now very walkable thanks to community efforts, regional environmental government assistance, and the organic industry of Townsville's nonhuman inhabitants. I'm torn. Herbert impresses upon me: this creek, like the creek in the novel's second chapter (entitled "psychological effect of a solar topee"), lies "like a mirror," reflecting not only what is viewing it and is above it, but also all other creeks that are full of life fading away in Capricornia and North Queensland (if not Australia) more generally. Instinctively, I notice something masculinist here, but I am a man with a predilection for metonymic distraction that keeps me from fixing political correctness to my emotions or even fixing my sun hat (taken out of desperation from the front of a campus store at the University of Queensland in Brisbane after my first journey into North Queensland). Mind and body aren't quite in control. To anchor myself, I turn to the passage "The creek lay like a mirror, fleckless but for chasings here and there." The narrative reaches towards an enigma. Something is fading deeply; an epic change is adumbrated by a "darkening"'in the "silence of the brooding bush": "Rich red gold was splashing on the waters of the reaches to the west, flowing to the sea in dazzling streams down gently-rolling troughs. The sun sank swiftly. Purple shade of night came creeping in. The red gold faded to the hard yellow gold coins, to the soft gold of flowers, to silver-gilt, to silver, to purple pewter chased with filaments of starlight."[18] Passing with the minutes, the narrator's sensual steps drop down in the color palette before human cries and a fire leap from a clearing, illuminating forms of naked men. The shadowed world, full of strange echoes and struggle under torches through the night, is one that is prefaced by the chastening lexicon of finance and greed: gold, coins, silver, pewter. The economy of the prose feels haunted in the way humidity snatches you from comfort or complacency, mobilized here to stop us from reading literally. Thus, proto-Anthropocene allegory accumulates through realistic detail in a similar fashion to the way Herbert's *Poor Fellow My Country* provides the reader with the precise notes and the exact timing of a change in octave to an Indigenous song in the haunting of a boy out fishing in the epic opening passage.[19] The novel is technology enough to afford the right spaces to gather these details and to allow such efforts to connect to the privatized emotions of the reader (magnetically, as it were). Rather than reaching for a direct apocalyptic register following the accumulation of facts, the narrator shepherds the micro and the macro into a voice born out of a prose style at ease with the discordant harmony of distorted nature writing and its particular musical pathos. It isn't a parable, but this passage speaks keenly to an epistemological emphasis that drives Australian fiction and its criticism of this period (and Herbert's is often driven by Indigenous knowledge, even though the text frequently rests at the edge of racism). There is, however, also the starlight that takes residence in the passage's final breath. That the unfathomably distant dying suns bring such illumination of space to us via the text is something to be reckoned with, wherever we are.

Countersignature 2

As I move out of the sun and take up the gently twisting path to First Avenue, Herbert's narrative structure fades from my mind and Stow's text emerges in my inner ear. More youthfully, I feel, I notice that under Marx's paws, leaves and nuts fallen from the soursop and native peanut trees crunch and pop. The sky, glowing light blue, reminds me of cyan (generated on a Z80 computer in the 1980s); it's fractured by chalky-white barks of the glorious foxtail palms (thrusting forwards from the wattles here and one of two rain trees that embrace the park on either side) rushing over and behind you to no end. More crocodile-warning signs, a small colony or copse of blue gums, and a white-bellied sea eagle populate the riparian revegetation whose Landcare heritage signage marks decades-long voluntary labor of the human immigrant populace that has transformed the space.

Tzvetan Todorov suggests that narratives operate on two levels: story (character and action) and discourse.[20] If we develop our criticism of Anthropocene autobiography by focusing on tenses, aspects, and modes of narration (discourse) that invoke the current and future trauma of air, land, and sea, we might begin to occupy fusions of poetics and history with what Rose terms "howling into, and from within the shadow place."[21] Out of this place of critical disobedience, I mark subjectively tangible experience of places as affordances for readerly empathy and grief. Like Mark in *Capricornia*, I'm close to trade winds, half-caste girls, doctors; unlike him, I am beginning, *with realizing*, "to think of these things as pleasant things, part of the pleasant world of which it [is] good to be a functioning part"[22]—although the irony that steels this passage (fleshed out in the shabby, unshaven slovenliness of this sexist character returning to sanity as a physical wreck) portends that the man will remain a stranger to this place, and subsequently the fortunes of human-environment fidelity might never be his. I see myself and I deny seeing him. One might compare two passages of chapter seven ("clothes make a man"), which demonstrates the ill-fitting of *anthropos* and *oikos*, the shift between dry season and wet season: "multi-coloured schisty rocks split with golden waterfalls" and "the harsh winds rioted with choking dust and the billabongs … mere muddy holes where cattle pawed for water."[23] Such difference might not be subject to simple contrast, for it is concerned with acting through opposing forces, oftentimes couched in the story of an exhausted, staggering search for meaning. The protagonist, as do I, embodies this quest.

Back in the office, my eyes catch a change to the light in the window behind my screen. A shape said medium-sized bird. The gregarious blue-faced honeyeater. His return suggests to me that my window is not "my window" but a long-standing sector of his habitat, formed for purpose by what he can see, smell, and hear here. I didn't choose to write about him; he came to the window. In this contact zone, his sensory values impress upon mine. Black, white, golden olive green; it's sometimes known as the bananabird in these parts, which have a history of slavery in banana and sugar cane plantations on land soaked in massacres of Indigenous Australians; one might footnote the massacres of European pastoralists, too. Systemic extermination. Defensive reprisals. Violence is violence is violence. How hard it is to shift from those realities of death to write something as superficial as my aching doubt. I'll never become much in the visiting bird's story, but what (over time) might he become in mine: sacred object, symbol of power, marker of transcendence or planetary grief? My two novels are perched on top of the system unit to the right of the monitor with half a dozen critical studies and notes from the

Figure 6. The *Monthly*, April-May 2019. Image copyright the author.

Figure 7. Black-Throated Finch. Flickr, Steve Dew. CC BY 2.0. https://www.flickr.com/photos/107368995@N04/33471514488/

archive of the Foundation for Australian Literary Studies. Through the forest of analysis, they are looking at me. The honeyeater walks along the concrete shades bisecting the first-floor windows along the social sciences building, his shadow cast onto The *Monthly* sitting on my desk (Figure 6); James Bradley's article on the controversial Carmichael coal mine and the black-throated finch of North Queensland (Figure 7) is marked for referencing.

Some weeks after writing these passages above, #Qexit and #MakeThemLeave started to trend. Following the results of the federal election, Australians were calling for Queensland to be removed from Australia. The failure was not with Queenslanders, but those who had to make the case for transitioning to zero emissions within a green economy that delivers more jobs for longer. It didn't happen. And the green vote held here, across the country, and grew 3.7% in the Senate.[24] But what was it all about?

The US $2 billion Adani investment project was launched in 2010, and Townsville residents have seen multiple announcements of an imminent start to the project since 2015.[25] It was made to seem that people wanted things to come to a head. Despite a whole suite of issues—the disappearing demand for coal in developed countries; China and India increasing import substitution; financial institutions under pressure to reduce their investments in financing or insuring miners, ports, and generators for coal; an increase in competition from gas and renewables since the project was first envisioned; the price of thermal coal dropping to US$90 per ton, which is US $20 per ton less than needed for the mine to be viable[26]—Adani seems set on adjusting its plans each step of the way to keep the project alive. With feasibility diminishing each quarter, the chief executive officer of Adani Australia, Lucas Dow, underscored the resource extractor's determination to push forward with the project even in the context of a diminished but ever publicly discernible compliance culture: "We are not going to be pig-headed about it"[27] (Figures 8–10).

In a suitably balanced article sensitive to public emotions, Amanda Cahill believes that the "rush to renewables" leaves communities behind and that regional communities are "reluctant to sacrifice existing

Figure 8. #Pastoral #StopAdani, Doongmabulla springs, Galilee Basin. Flickr, Tom Jefferson. CC BY 2.0 https://www.flickr.com/photos/148888193@N07/33405744582

Figure 9. #pig head. Christopher Gillette, ABC News.

Figure 10. Deirdre Boeyen-Carmichael, *We will not be silent*, 2019, oil on paper.

industries that have jobs available here and now in favour of investing in new projects."[28] From a science-writing perspective, Bradley alludes to the bag of dirty tricks up Adani's sleeve: playing it tough in the courts, particularly with proceedings against local Indigenous groups.[29] As a counterpoint, Bradley cites one of three spokespeople for the Wangan and

Jagalingou people who value the Doongmabulla as a site of creation within their cosmogony: "its waters are the place from which their creator spirit, Mundunjudra, travelled through the land, giving shape to it and defining the system of totems and rituals that connect the traditional owners to their country."[30] At this point, mass protests gathered along the Queensland coast and a convoy of activists moved through the country from Hobart to the Galilee Basin (Figure 10) and on to Parliament in Canberra. To sum up, illiteracy for sacrifice and bullying spill over into contested senses of spirit and protest, which both bind and rupture the community alike.

Anthropocene Field Notes 3

The black-throated finch looked unstoppable in April—unstoppable as a news story, that is, not a species. In the lead-up to the 2019 election, the finch hit the newspaper front pages across Queensland like no other bird on the island-continent this century. It didn't know it but the estrildid finch dominated leader commentary and letter pages for months as the polarizing debate about its decline came to life in light of the protracted delay to Adani's Carmichael coal mine in the Galilee Basin. For good reason.

I thought I'd pick up a local paper while grabbing a coffee as I was exploring the remains of the train line and its heritage signs on my first day in Townsville. Full-page spread on a little bird!? I was reminded of my very first day on arrival in Australia. Sitting under the Sydney Harbour Bridge in May 2011 with a train ticket to Armidale in my jacket pocket, I turned to the free airport copy of the *Australian* to find a two-page spread about an Indigenous group selling land to developers on the way to securing a health-care center. I didn't need to suspend disbelief; I had to withhold judgement. I need to understand this country through its practices, values, and discourse, not through my own. Years later, the image of the bird and the raw denialism in the letters page of the *Townsville Bulletin* offered another moment to exercise restraint. Ornithographers confirmed observed decline in extent of occurrence of up to 59% in the last decade from Cape York Peninsula to Queensland, yet none of the editorials I could see mentioned that the species was listed as endangered in September 2017,[31] nor that it was listed as extinct in February 2019.[32] As with the text I read under Australia's iconic soaring steel arches, there was something missing from the narrative. I started to scratch the surface and found that the black-throated finch has been absent from Brisbane since the 1930s, but it is locally known to be found around Townsville (which hosts Adani's regional headquarters) and

Charters Towers, and between the Burdekin and Lynd Rivers. So, it's here, but it might not be for long. I look for it as I go walkabout my way. I wonder, as with the loss of so many species to the extractive industries around the world, whether the finch will become the subject of elegy, lament, or the politics of impotence here in Queensland soon.

Nostalgia is a scandal to Anthropocene ecomodernists but, as with elegy, it will go on, only not quite the same; the practice meets a new cultural need as species after keystone species pass. In the time of the breaking of our planetary carrying capacity, plants are on the move. Animals are thinning out. For all the entanglement of species we wish to promote, one thing separates us from them most poignantly: our signal hope is that these Anthropocene annals cloud into night before the songs of the birds die.

Bradley's article brings us back to countersignature and the devastating and complex nature of being on track for planetary heating of somewhere between 3 °C and 4 °C by 2100: "the juxtaposition of the Reef and Carmichael … crystalized disquiet about the wisdom of opening new coalmines at a time when the effects of climate change were ever more obvious."[33] I'm trying to see things in conflict while also opening myself up to the dialectical relationship between our world and the world to come.[34] It isn't easy for me to read Karen Barad right now, but I think it might help if I'm opening up to an agential realist ontology. An informed stance enables us to notice the world as not composed of discrete "things"—that is, human, dog; bird, coal mine; pig head, protest—but of "phenomena in their becoming."[35] With the ontopoetics of the birds in mind, as Mick Smith notes, these phenomena are most often "multi-species" and of "multi-existent" communities of different beings, "each exposed to each other in myriad different ways."[36] It's relevant to Bradley and Townsville politics, but it needs something else.

Countersignature 3

I'm new to fictocriticism but I aim for creative exploration and the spontaneity of organic connection. Kerr notices something similar to my ambition, to remain open to the world's provision (as with Barad and Smith), in the accomplished aesthetic of Stephen Muecke's work. Muecke's "methodological parallelism" envelopes an ethical text *and* its commentary, whereby "a figure for metaphoric blurring (montage) is replaced by a figure for analogical contiguity."[37] This helps me work out what might happen in the experimental countersignatures in this article, and what we might add practically as writers to the advances of theoretical insight. The emphasis on contiguity trumps the emphasis on change or difference in environmental and emotional histories, for it places things that are clearly different on the surface (montage, medley, mixture) as potentially similar or connected within a wider emergence of life. To keep this dialogue open, my images provisionally mark out discovery and digression as they meet in a frenzied snatch at the keyboard on the way to publishing for a salary. I'm encouraging readers to look across worlds, and across each other and species; as with analogy, we come to conclude that some further similarity exists between things that an emphasis on difference betrays.

Some weeks ago, at my desk in a living room of a miller's cottage in the first European inland settlement of Australia (Bathurst, New South Wales), I wondered what Marx was thinking as he looked at me through the screen door: it's hot out here, a few tapping sounds and a bright light, Daddy's back to me, time for a walk (Figure 11). Weeks later, this momentary lapse of selfishness came back to me when I first sat in the delimited space of my office. I was typing up my notes on countersignatures as the blue-faced honeyeater appeared at my window; however, the fleeting companion species didn't trigger a thought towards all the new relationships I was about to experience in the tropics. It was a specific feeling that came back into life from a darkness within, triggering a

Figure 11. Marx, Bathurst, 2019. Image copyright the author.

Figure 12. Prometheus Chained, 2019. Image copyright the author.

memory of working through a metaphor of entrapment that I started in New South Wales. I had placed myself a short stroll from Cox's Road from the Blue Mountains to the Bathurst Plain (1814) to traverse physically the historical world of Lachlan Macquarie (and his resistance to power and unchecked pastoral growth) during yet another devastating drought in the home of the country's European pastoral expansion beyond Sydney. For two months, I synchronized my diary with his and began to write on the multiple fronts he mobilized to transition Australia from colony to country. But there was too much patriotic spin in all this. Now, critical anxiety came back to me at the desk by the window shortly after returning from viewing the most compelling representation of power and its restraint that I had seen so far this year (Figure 12): a figurative sculpture *Prometheus* (Jane Hawkins, 1997), demonstrably chained to the wall of the decolonized space that is the countersigned Mabo Library.

This was my first day at James Cook University. The compression of multiple sources of conflicting emotions impressed upon me for about ten seconds, freezing me at the keyboard. It was a moment you might name uncanny, syncretic, and hyperpersonal. If nothing else, it is part of my

Figure 13. Self-portrait at JCU, 2019. Image copyright the author.

autobiography, which instances how I feel living through the Anthropocene within one Australian institution while writing for colleagues in another (Figure 13). I finished that sentence and took the bus back to the Sovereign Hotel while editing a few words on situated availability.

Anthropocene Field Notes 4

We need to enlarge on Eakin's useful definition of "situated selves": discrete narrators who position themselves "as products of a particular time and place."[38] After working through the ideas present in the work of Donna Haraway and Doreen Massey, conflating one dichotomy (planet and human history) with another (self and world) might be wrongheaded, egotistical, and contrary to a poetics of what I might name, after Eakin and Rose, the open dialogue of identity-shaping environments (the acronym might be O-DIE). The natural countersignatories to our existence, autobiographical enterprises, can be, as Eakin understands, "nested one within the other—self, family, community"[39]—yet equally set in non-metaphorical spaces of more-than-human adaptation and loss.

Nesting is a good metaphor, but the black-throated finch in North Queensland, the School Strike for Climate kids and Extinction Rebellion youths, and refugee academics all across the world repeat the struggle of species worldwide. To keep with Eakin, it might be better to articulate the struggle for meaning (and a lack of settlement on meaning) in the conflicting juxtapositions we encounter more readily this decade as we fight through the polarizing prose of the press and its discontents, crystallized in the copy countersigning politics of care into coal versus coral in the *Townsville Bulletin*, for example. Moreover, the emphasis on human agency in Eakin is clearly discordant to those of us thrown into this mess more than we are empowered to work through it. Following the deletion of the hubristic liberal subject by ecological philosophy in the last century, we now need to check our figuration of agency and responsibility. If only I could write that last sentence without being human.

Postscript

I'm tempted to investigate the various ways writers of this island-continent countersign histories of life with hyperlocal lexicons while reaching out most broadly to our species' capacity to empathize with others; it is something often occluded by narratives and histories of violence. In Australia, we are particularly sensitive to the precariousness of survival. But right now, in North Queensland, this feels like a bourgeois thing to write. Death here, like elsewhere, is not only of nonhumans; it is of human languages and of the longest-standing human practices that were underpinned by the Great Australian Peace.[40] It is the site of war that bolstered an identity which continues to claim the right to interfere with land in a way that leads to human and nonhuman losses at scale. A few post-structuralist concepts bring little justice to either this history or to the challenge of reconciliation.

Rather than looking out for ethical, progressive, or radical ways to color a metanarrative of the relationships between individual species that make up the more-than-human world and its suffering, the absence of absolute boundaries might be worth exploring in the light of the contiguity and contagion that stereometry delineates. It might also help us delete the privatized human "I." In a world where nothing is final or absolute, we might more concretely understand things in all their movements—not only in their lifeworlds, ecologies, and cultures, but with their opposite and contradictory sides in unity. I have tried to point to this by foregrounding current destabilizing experiences in my life in this experimental work.

Inspiration for such writing came to us a century before environmental humanities and multispecies ethnographies landed in the academy. Friedrich Engels invited us to move in this direction in his introduction to *Dialectics of Nature*: "the life of being conscious of nature and of themselves, is just as narrowly restricted as the space in which life and self-consciousness come into operation. A cycle in which every finite mode of existence of matter, whether it be sun or nebular vapour, single animal or genus of animals, chemical combination or dissociation, is equally transient, and wherein nothing is eternal but eternally changing, eternally moving matter and the laws according to which it moves and changes."[41] Narrowly restricted spaces haunt this piece of writing and its ethical imperative to escape the confines of genre. Todorov explains, however, that genre isn't narrow: "Genres are the meeting place between general poetics and event-based literary history; as such they constitute a privileged object that may well deserve to be the principle figure in literary studies."[42]

I end this piece by bringing together two pieces of Australian fiction, drawn from opposite sides of the island-continent to animate autobiography as a privileged object in these terms. Both authors exemplify a bioregional imaginary in the wake of colonization. For my study of literature in Australia, I feel that I need to process the novel's capacity to hold open unprecedented loss, and to analyze the ways it feels out the shadowland we inhabit owing to violent external forces and those of human making. To complete Engels' project, one of hypersensitivity to motion, we must return to the novel's movement through geographic space, its turning towards and away from change as registered in the forms and modes that speak most broadly to our emotions. We might also stay put and listen to country.

These ideas placed in opposition didn't sit well with my pursuit for open dialogue. This bothered me while I was wondering whether I could approach Stow and Herbert together; I felt that such a move would require an obscure intellectual abstraction (external violence) and a leap of the imagination (to region), which made me feel unhappy about it all. I decided to set out to clear my head and ease the dissatisfaction. I thought I'd follow the railway line out of town towards the most flood-prone area to see if I could discern the damage from the summer event that swept away hundreds of cattle along with the stations. On my way, there was something flashing in the light of a flood scar on the opposite side of the road. I pulled up. Was it a grave? (Figures 14 and 15).

It was during this period of reading Stow and Herbert together that I discovered Townsville has a history of Japanese air raids. After speaking with my colleague back in the office, I was put in a mild state of shock,

Figure 14. Oonoonba, July 2019. Image copyright the author.

Figure 15. Third Japanese raid on Townsville, Qld - WW2. Flickr, Aussie Mobs. Public Domain Mark 1.0.

realizing that on my drive out of town I had stumbled across the final remains of the evidence of a raid that hit an agricultural research station in 1942. A fading scar, barely discernible now as this summer's environmental disaster washed out its presence. Like the two male novelists, the flood had taken me to the last century's World War II. So what follows is relevant, if only as a part of my autobiography as I turn to the Australian novel:

> The hut was pitch-dark, steam-hot. It stank of men and the tropics.
>
> The man by the wall, whose eyes were accustomed to the hut, looked at the shape of the man who had fallen beside him. After a while he said: "Okay?"

> The man on the floor said nothing. His mouth was in the dirt.
>
> "You all right?" whispered the man by the wall. He reached out to the other man. "Hey, mate."
>
> The man on the floor said nothing.
>
> "That's blood," said the man by the wall. "Hey. What did they do? Are you hurt?"
>
> The voice of the other man came out of the dirt muffled. "Not mine," he said. "Not mine."
>
> "What happened?? Are you hurt?"
>
> "I'll live."
>
> "The blood———"
>
> "I don't know his name," said the man on the floor. "We were tied together. They cut his head off."[43]

> A path of light led from the steamer's bow to the Japanese cruiser, which lay at anchor in midstream, a huge black mass against the starry southern sky, raking the Southern Cross down with her alien rigging. … Many thousands of miles away as the southern crow would fly should he wish to travel to another hemisphere, the earth and heavens were being rent with violence scarce imaginable, and men were being smothered, burnt, and torn to death, while here, in spite of the machinations of murderous propagandists in a foreign land, the tide was whispering to the ancient shore, snipe were playing on the beaches, hummingbirds were sipping honey from the heats of flowers. What a world it was in the hands of men![44]

To elect two moments of Australian literature that are cast in the light of Australian-Japanese relations either side of the war, in an attempt to secure a sense of nation as set in narrative, is nebulous at best. I've put them here to speak of literature's unparalleled capacity to harness and manipulate feeling as I come to write out Anthropocene autobiography. To think of these works now and in the light of the scholars I've placed before us in this article, we might attend to the ways human voices modulate as the world lit by the torch of violence comes into presence. Here, it forms consciousness in distinct encounters: direct speech in the moment of accounting for the bloodshed in Malaya for peacekeeping services defined in another part of the globe (Stow) and the wisdom and poetry of the earth-inflected narrative voice slowly building into apostrophe after carefully piecing together distinct and distant elements of the world reeling into its decline (Herbert). We have much to learn from Australian fiction.

Amitav Gosh suggests that the climate crisis requires "a departure from our accustomed logocentricism."[45] Fictocriticism might enhance autobiographical writing by connecting life writing to new confessional cycles of self-consciousness that seem not only to come and go with the climate, but also to get trapped in the carbon sink of the novel; like ice cores, these can be unplugged from the archive and examined for evidence of who we are and who we have been. Whether this will offer much greater purchase than the logocentrism of history alone, time, like the moon (tired these million years of the arguments of humans), will tell.

James Cook University

Disclosure Statement

No potential conflict of interest was reported by the author(s).

Notes

1. Herbert, *Capricornia*, 23.
2. The Australian Government Bureau of Meteorology lists twenty-one cyclones on Queensland's east coast since 1890, with seven occurring since 2000.
3. Eakin, "Autobiography," 271.
4. Eakin, "Autobiography," 273.
5. See the epigraph above.
6. Derrida, *Limited Inc.*, 55, 78.
7. Wills, "Derrida and Aesthetics," 126.
8. Barthes, "The Photographic Message," 17.
9. Barthes, "The Photographic Message," 17.
10. Bilborough, "Poetic Representation," 182.
11. Rose, "Slowly," 3, 5.
12. Kerr, "Fictocritical Empathy," 194.
13. Rose, "Slowly," 1.
14. Lobb, *The Flight of Birds*, 217.
15. Lobb, *The Flight of Birds*, 21.
16. Stow, *The Merry Go Round*, 56.
17. Herbert, *Capricornia*, 15.
18. Herbert, *Capricornia*, 15.
19. Herbert, *Poor Fellow My Country*.
20. Todorov, "Structural Analysis of Narrative."
21. Rose, "Slowly," 1.
22. Herbert, *Capricornia*, 27.
23. Herbert, *Capricornia*, 68, 69.
24. Denniss, "The Morrisson Election."
25. At the time of writing, Adani had to upgrade its environmental plan for the finch, which was limited to the site of a pastoral lease (with continual cattle grazing) at Moray Downs west. Further to this, Adani needed to provide a separate groundwater plan to manage groundwater-dependent ecosystems

and definitively identify the source aquifers of the Doongmabulla Springs, regarded as one of the world's last pristine desert oases.

26. See Bradley, "Holding on to Madness"; and Quiggin, "Adani's Carmichael Mine."
27. Hanman, et al., "'Not pig-headed': Adani emerges confident after tense coal mine talks."
28. Cahill, "Don't Condemn Voters."
29. Adani has commenced proceedings against Adrian Burragubba (to the tune of AUD $600,000).
30. Bradley, "Holding on to Madness," 32.
31. See Queensland Government, "Nature Conservation Act 1992."
32. See New South Wales Government, "Biodiversity Conservation Act 2016."
33. Bradley, "Holding on to Madness," 30.
34. Following Rose, "Slowly," 3, above.
35. Barad, "Nature's Queer Performativity," 46.
36. Smith, "Ecological Community," 30.
37. Kerr, "Fictocritical Empathy," 185.
38. Eakin, *How Our Lives*, 85.
39. Eakin, *How Our Lives*, 85.
40. See Pascoe, *Dark Emu*.
41. Engels, *Dialectics of Nature*, 54.
42. Todorov, "The Origin of Genres," 201.
43. Stow, *The Merry Go Round*, 42.
44. Herbert, *Capricornia*, 117.
45. Gosh, *The Great Derangement*, 83.

Acknowledgments

This work was made possible through the European Union's Marie Sklodowska-Curie Actions COFUND scholarship program and the Colin and Margaret Roderick Trust, Australia. I wish to thank Jessica White and Gillian Whitlock for their encouraging and authoritative responses to this article. I also wish to underline my gratitude to Mark Rawlinson (University of Leicester), former colleagues at the Australian Research Council's Centre of Excellence for the History of Emotions (University of Melbourne), and to Claire Brennan and Richard Nile (James Cook University). This article is written in memory of Dr. Heather Kerr.

Works Cited

Barad, Karen. "Nature's Queer Performativity." *Women, Gender and Research* 12, no. 1–2 (2012): 25–53.

Barthes, Roland. "The Photographic Message." In *Roland Barthes: Image–Music–Text*, edited by Stephen Heath, 15–31. London: Fontana, 1974.

Bilborough, Paoloa. "Poetic Representation as Social Responsibility in a Collaborative Documentary." *a/b: Auto/Biography Studies* 28, no. 2 (2013): 180–191.

Bradley, James. "Holding on to Madness: How an Infatuation is Keeping Adani's Mine Alive." *Monthly*, Apr-May (2019): 19–39.

Cahill, Amanda. "Don't Condemn Voters Who Fear the Transition to a Greener Economy." *Guardian Weekly* 31 (2019): 28.

Denniss, Richard. "The Morrisson Election: What We Know Now." *Monthly*, June (2019).

Derrida, Jacques. *Limited Inc.* Translated and edited by Samuel Weber. Evanston, IL: Northwestern UP, 1972.

Eakin, Paul John. "Does Autobiography Have a Future?" *a/b: Auto/Biography Studies* 32, no. 2 (2017): 271–273.

——. *How Our Lives Become Stories: Making Selves.* Ithaca, NY: Cornell UP, 1999.

Engels, Friedrich. *Dialectics of Nature.* London: Lawrence and Wishart, 1946.

Gosh, Amitav. *The Great Derangement: Climate Change and the Unthinkable.* Chicago, IL: U of Chicago P, 2017.

Hanman, Peter, Cole Latimer, Tony Moore and Eryk Bagshaw, "'Not pig-headed': Adani emerges confident after tense coal mine talks." *The Sydney Morning Herald*, 23 May (2019).

Herbert, Xavier. *Capricornia.* Sydney, NSW: Angus and Robertson, 1938.

—— *Poor Fellow My Country.* Sydney, NSW: Angus and Robertson, 1938.

Kerr, Heather. "Fictocritical Empathy and the Work of Mourning." *Cultural Studies Review* 9, no. 1 (2003): 180–200.

Lobb, Joshua. *The Flight of Birds: A Novel in Twelve Stories.* Sydney, NSW: Sydney UP, 2019.

New South Wales Government. "Biodiversity Conservation Act 2016." Accessed November 7, 2019. https://www.legislation.nsw.gov.au/#/view/act/2016/63

Pascoe, Bruce. *Dark Emu: Black Seeds: Agriculture or Accident.* Broome, WA: Magabala, 2013.

Queensland Government. "Nature Conservation Act 1992." Accessed November 7, 2019. https://www.legislation.qld.gov.au/view/html/inforce/current/act-1992-020

Quiggin, John. "Adani's Carmichael Mine Is Unlikely to Go Ahead, and Most People Know It." *Guardian*, May 3, 2019. https://www.theguardian.com/commentisfree/2019/may/03/adanis-carmichael-mine-is-unlikely-to-go-ahead-and-most-people-know-it

Rose, Deborah Bird. "Slowly—Writing into the Anthropocene." *TEXT* 20 (2013): 1–14.

Sebald, W. G. *Austerlitz.* Munich: Hanser, 2001.

Smith, Mick. "Ecological Community, the Sense of the World, and Senseless Extinction." *Environmental Humanities* 2, no. 1 (2013): 21–41.

Stow, Randolph. *The Merry Go Round in the Sea.* Harmondsworth: Penguin, 1968.

Todorov, Tzvetan. "The Origin of Genres." In *Modern Genre Theory*, edited by David Duff, 159–170. New York, NY: Longman, 2000.

——. "Structural Analysis of Narrative." *Novel* 1, no. 3 (1969): 70–76.

Wills, David. "Derrida and Aesthetics: Lemming (Reframing the Abyss)." In *Jacques Derrida and the Humanities: A Critical Reader*, edited by Tom Cohen, 108–131. Cambridge: Cambridge UP, 2001.

Writing Toward and With: Ethological Poetics and Nonhuman Lives

By Stuart Cooke

ABSTRACT
In this essay, the author argues that the appreciation of nonhuman poetic forms, or an "ethological poetics," is a necessary but neglected mode of ecological relation, and is especially important in the Anthropocene. Motivated by his own creative practice—in particular, the composition of *Lyre*, a book of poems about different animals, plants, and landforms—he considers important examples of ethologically attentive poetics before outlining how his compositional method attempts to incorporate insights from the environmental humanities and animal studies. Rather than insisting on their essential difference from human worlds, the author argues for an attentive, ethical, and imaginative engagement with nonhuman lives, through which surprising and unusual forms of poetry might emerge.

The Anthropocene is, first and foremost, about death on an unimaginable scale. Human-induced climate change and habitat destruction have wiped out sixty percent of mammal, bird, fish, and reptile populations since 1970,[1] and we are currently causing anything from two hundred to ten thousand extinctions each year.[2] As exploding human populations bring us into ever closer proximity with dwindling populations of other species, and as our industrialized cultures become ever more inextricable from other species' extinctions, the need to acknowledge the importance of nonhumans in human worlds is becoming increasingly pressing. In this context, understanding nonhuman poetics "helps us to see the Earth as something else than we often do: a place with other life, other presence, other beauty."[3] In turn, we might develop an awareness of what Kate Rigby calls "the semiosis of the more-than-human world," in an effort to overcome a "perilous condition of self-enclosure" that renders industrialized cultures "dangerously oblivious" to their relations with other creatures.[4] In this essay, I outline a response to such more-than-human semiosis. I refer to "ethological poetics," or the study of nonhuman poetic forms, and in particular to their influence on my own poetry. For the past seven years, I have been writing a collection of poems that attempts to imagine the lifeworlds of nonhuman animals, plants, and landforms such as volcanos and estuaries. While the project began with a relatively simple objective—to respond, as a poet, to the songs of other species—my motivations were various. On the one hand, I was interested in rescuing nonhuman subjects from the backdrops of human dramas, in creating art that "puts into place a creative interruption of the ways in which humans habitually look at animals."[5] On the other hand, and perhaps less nobly, I was also interested in writing compelling and original poetry. Les Murray's 1992 collection, *Translations from the Natural World*, remains one of my favorite books, not least for how its cross-species explorations lead the poet into some of his most experimental and linguistically remarkable poetry. What for many critics was an unfortunate deviation from Murray's more conventional poetics, *Translations* remains for me an indelible example of my argument here: that multispecies collaborations can produce the most incredible events.

Ethically and aesthetically, then, I had a sense that an ethological poetics was important both for the planet and for my poetry. However, in attempting to respond to the songs of other species and, where possible, to translate their calls into poems, I soon realized that "songs" were only the most sonorous of the vast array of nonhuman expressive forms. While I started with a relatively naïve interest in bird songs, my research very quickly began to engage with the implications of biosemiotics and speculative forms of ethology. Inspired by Vicki Kirby's expanded understanding of Jacques Derrida's "open system," in which literate expression is dispersed not only across human forms, but also throughout "the weaving of life itself," I was confronted with the possibility that I could look for expressive examples of textuality not only in other animals, but also in all manner of organic and inorganic forms. For Kirby, even a photon can become a subject of "cognitive and agential entanglement and observational intention."[6] If everything, in these terms, can be considered to be "alive," it is not because everything is an organism or organized as such; rather, as Gilles Deleuze points out, the organism is one of many possible diversions of life itself.[7] I was and remain excited by symbiosis and synthesis, as opposed to enforcing categorical distinctions and hierarchies of evaluation.

I am also aware, however, that while various links between art and biology have been made, willingness to assign real correspondence between human culture and the rest of the biosphere, to quote James

Sherry, "has been partial and threatening to most [Western] intellectual frameworks."[8] It is one thing, therefore, to follow the important work of people like Isabelle Stengers and Deborah Bird Rose in reclaiming symbiosis and mutualism, rather than cutthroat competition and differentiation, as "utterly fundamental to life on earth."[9] But it is something else entirely to propose that such symbiotic relations might find ground in poetics—in other words, that aesthetics might actually be a more-than-human concern. As Timothy Morton argues, however, aesthetics is absolutely central to animal politics: "One of the deepest questions is not, 'Can animals think?' but 'Are animals capable of aesthetic contemplation?' Why? This would mean that the aesthetic was not a 'high' function of 'greater' cognitive powers, but a 'low' one. What if the aesthetic were the default mode of sentience as such?"[10]

As I have proposed elsewhere, an "ethological poetics" refers to critical methodologies for the reading of nonhuman art.[11] By studying examples of creative expression in nonhumans, ethological poetics deconstructs anthropocentric conceptions of "art," "culture," and "language." Of course, such an endeavor cannot escape important questions about aesthetic categories (Is a bird's song really a song, for example?), but such questions are not necessarily the *right* ones (after Vinciane Despret). Beyond concerns of categorization, the important point is that when we consider compositions as complex as, say, those of the Albert's lyrebird or the humpback whale, we are compelled to extend the bounds of art—in terms of both what it is and where it comes from. In doing so, we are not only bringing the realm of aesthetics "down," in Morton's terms, from a transcendent realm of human cognition but, perhaps more importantly, we are becoming attentive to other creatures' communities, and to their desires and aspirations. To direct attention to the umwelten of our nonhuman companions is the objective of an ethological poetics, therefore, which follows and then departs from the now common "*eco*[logical] poetics." Ecopoetics tends to refer to human poetry about broader environmental processes, or to those environments *in which* organisms live and die. Environments are not generic containers into which organisms can be placed, however, but are created and shaped by these organisms.[12] As Morton writes, "the idea of 'our' environment becomes especially tricky when it starts to slither, swim, and lurch toward us."[13] In a way, then, by sidestepping concern with "the natural world," ethological poetics responds to Morton's claim that *ideas* of nature, environment, and so on are "getting in the way of properly ecological forms of culture, philosophy, politics, and art."[14] An ethological poetics directs thought toward organisms, however these may be conceived, and their myriad signals that we cannot immediately see or hear or understand. As a politics, ethological poetics corresponds with what Morton would call a "leftist ecology": opposed to fascist conceptions of "a greater whole," ethological practice resides "in the singularity of, and conscious commitment to, the other."[15] Here, of course, ethics is no longer concerned with the correct response to a radical alterity, but with how we respond to and care for "the lively relationalities of becoming of which we are a part."[16]

As Jonathan Balcombe writes, "[t]he question is no longer Do animals think? but What do animals think?"[17] Nevertheless, Despret argues that it remains the case that animals are suspected much more often than humans of lacking autonomy, particularly "when it concerns actions, such as *cultural* behaviours, that have for a long time been considered as proper to man."[18] This situation is particularly acute in anglophone literature, in terms of both its production and its criticism. While innovative eco and animal art has an established history in the West, literature lags well behind. For at least the past 250 years, Sherry points out that much of the Western literary avant-garde has focused on "social life in cities," which "obscures the truism that human activity is not unnatural or separate from nature."[19] According to Graham Huggan and Helen Tiffin, the reasons are various: "literature, with its traditional emphasis on plot, character and psychological states, has been seen perforce as being focused on individuals or groups of humans, or at least anthropomorphised animals. … And while literature has certainly dealt with the fates and even the psychologies of animals, these have—at least until recently—been highly anthropomorphised, acting more often than not as a staple of fiction for children rather than adult readers. For western writers, at least, it has been more difficult to anthropomorphise the environment which … has been regarded as a mere backdrop against which human lives are played out. And even when writers have given some attention to the natural (extra-human) environment, critics have generally downplayed its significance in their own considerations of the work."[20]

Clearly, Huggan and Tiffin are focused more on narrative-based genres such as the novel and the short story than they are on poetry. Because it need not rely so squarely on human plot or psychology, however, poetry arguably has been more open to the possibilities of nonhuman semiosis. There is also a more obvious formal correspondence between the musical heritage of poetic genres such as the lyric and, for example, bird songs. Indeed, many attempts to describe bird song with standard musical notation have resulted in the researcher being "driven into poetry" instead.[21] A notable early example of such poetry is Walter Garstang's *Songs of the Birds*, which collates his "nuanced transcriptions of bird songs" as a series of sound poems and lyrics.[22] Notable both for Garstang's "rigorous, exact transcription" of the structure and timbre of avian songs and the way in which he "stretch[es] language towards the limits of syntax," *Songs of the Birds* is a fitting testament to how careful observation of the nonhuman can tend toward a kind of poetry, and of how that poetry is invariably highly original and surprising.[23] Certainly, the situation has changed dramatically in more recent decades, with Murray's and Garstang's titles but parts of an array of remarkable works of both fiction and poetry, which also includes Barbara Gowdy's *The White Bone*,[24] Laline Paull's *The Bees*,[25] and Gianni Siccardi's *The Blackbird*.[26] Nevertheless, it remains the case that when we compare this array with the abundance of human-centered literature, it is clear that there is a relative paucity of fiction and poetry that seek to venture beyond superficial description of nonhumans, and to rigorously observe and articulate their ontologies. The lacuna suggests that language and the language arts remain what Donna Haraway would call one of those "last beachheads of human uniqueness."[27]

In large part, this failure to imagine nonhuman others is indicative of the legacy of European colonization and hegemony: colonialist classification and interpretation "necessarily resulted in the destruction or erosion of alternative apprehensions of animals and environment, blocking understandings of those crucial interactions between the human and the 'extra-human.'"[28] A sign of neocolonialism in much contemporary thought is the pervasive fear of anthropomorphism, which even rears its head in the larger quote from Huggan and Tiffin above. While the filmed, photographed, or painted animal can remain enframed and, therefore, at a distance, it seems that the animal who speaks in language inflames that very prejudice which many of us, otherwise, would condemn: that only humans can speak of things and know of them as things; that animals, in Martin Heidegger's terms, are "poor in world";[29] and that an animal language is only a product of human imagination. But, to be blunt, these claims have little interest for me. With Val Plumwood, my feeling is that the charge of anthropomorphism too often shuts down discussion, rather than opening up inquiry about compelling instances of cross-species relation.[30] As Aaron Moe has put it, anthropomorphism is only a "fallacy" when one ignores the innumerable, compelling continuities between "human" and "animal" spheres.[31] While, of course, we can reject "the facile and basically imperialist … move of claiming to see from the point of view of the other," Haraway reminds us that "the philosophic and literary conceit that all we have is representations and no access to what animals think and feel is wrong."[32] Disregarding the fantasy of climbing into one another's heads, we can still make "some multispecies semiotic progress" based on all manner of ethological, biological, and neurological evidence. To refuse, or "[t]o claim not to be able to communicate with and to know one another and other critters, however imperfectly, is a denial of mortal entanglements … for which we are responsible and in which we respond."[33]

To be sure, ethological poetics should be situated within a tradition of decolonial activism and scholarship. After centuries of writing *about* the experiences of Indigenous peoples, and of confining scholarly discussion of their arts to the realms of archaeology and classical anthropology, recognition and appreciation of Indigenous cultural expression has grown exponentially in the past half-century. In Australia, North America, and a range of Latin American countries, we are well aware of the immense value of burgeoning Indigenous publishing cultures—in part, because of the sophistication and beauty of Indigenous literatures and, in part, because these literatures are such powerful means of resistance against imperialism. An ethological poetics attends to the *other* creatures whose lives have been altered irrevocably by European invasions; *terra nullius*—a land with no people or voice—becomes a thriving cacophony of human and more-than-human expression. After all, a rigorous decolonialism should involve a careful and ongoing consideration of First Nations' cosmological systems. Invariably in such cosmologies, nonhuman animals are as integral to conceptualizations of "land," "country," and "culture" as people themselves. In Australia, for example, proper recognition of Aboriginal sovereignty implies recognition that other animals and plants also have their own law. Consider, for example, the following from Rose: "According to my Aboriginal teachers, plants as well as animals are sentient, and the earth itself has culture and power within it … we are all culture-creatures, we are intelligent, we act with purpose, we communicate and take notice, we participate in a world of multiple purposes. It is a multi-cultural world from inside the earth right on through."[34] In their efforts to unravel the legacies of the colonial imagination, poets and scholars should think very carefully about the implications of First Nations' ontoecological terrain.

As a poet, my creative practice takes this challenge—to pursue multispecies entanglements within a literary milieu that is still largely afraid of dealing with them—as a source of inspiration. For Derrida, after all, "thinking concerning the animal … derives from poetry." Poetic thought constitutes "what philosophy has, essentially, had to deprive itself of."[35] Situated at the limits of philosophy, the poem necessarily contends with the limits of human cognition as well. Indeed, my interest in ethological poetics stems from the direction that my poetry has taken over the past decade: in seeking to articulate more-than-human umwelten, I have been confronted repeatedly with the fact that my own mode of expression is but one of many in a much broader ecological array. A writer in this position quickly becomes preoccupied with finding "language that coheres with evolution, with our destiny as animals among other plants and animals."[36] For human language to "cohere" with a more-than-human earth, the process of *translation* is of paramount importance; it is no accident, for example, that Murray's collection is named Translations *from the Natural World.* Rather than enacting dangerous leaps of imaginative imperialism, such poetry is a thoroughly necessary mode of ecological relation. As a form of ethological enquiry, it can take language and, with it, imagination to the very edges of human perception; here, in this open, indeterminate region, what we think we know might mingle and even start to merge with what is thoroughly unfamiliar. The unfamiliar is vital because the "coherence" of language and earth cannot become a domination of the former over the latter: the translation must produce a sense of the earth's otherness, rather than attempt to harness its agency; the rules of language must be stretched rather than religiously obeyed. I make these claims in the interests not only of good ethics but also of good poetics: as much as language is a cognitively rich, phenomenological process, following Martin Harrison, I argue that there is no immediate symbiosis between language and thought. Rather, writing produces an inevitable, and necessary, gap between thought and the written composition, a gap which is at the heart of poetic effect. A good poem, therefore, is driven toward both construction *and* deconstruction. Without this interplay between assemblage and disassemblage, between a work's apparent coherence and its capacity always to slip beyond the bounds of any determination, concept and sensation are subsumed in one another. Thus, language will revert to being a merely representationalist medium and the constructive, ongoing, and compelling energy of poetic meaning will be weakened.[37] As the poem is weakened, so too is the potential for us to imagine its subject. At its best, as a compressed form in which language is at "its most agile and expressive,"[38] poetry provides the potential for meaning, sound, and (typographic) design to weave, like the nonhumans that surround us, ceaselessly in and out of our normative fields of comprehension.

I mentioned earlier that multispecies collaborations often produce extraordinary results. Indeed, the focus of Moe's *Zoopoetics*, is on how the attentiveness to nonhuman species has produced remarkable formal and stylistic innovations in North American poetry for more than a century. The presence of the nonhuman in such work need not be made explicit, or its language translated in full. In Wendy Burk's *Tree Talks*, for example, the poet's encounters with individual trees produces a growing absence at the heart of the book, as it becomes increasingly apparent that, indeed, trees *do not* talk. However, rather than nullifying their presence in an imposed silence, the poet renders the possibility of talking trees more *likely*. Structured as a series of interviews with particular trees across Southern Arizona, Burk asks each of her arboreal subjects a set of starkly prosaic questions. "Do the dogs bark at all times of the day?" she asks a velvet mesquite in Tucson. "Do you get used to it?"[39] Then, she provides elaborate transcriptions of the trees' responses. Interestingly, Burk's transcriptions account for all kinds of peripheral noise, human and otherwise, while the trees themselves remain tantalizingly on the edge of written language, their responses rendered non-alphabetically with different arrangements of diacritical marks such as slashes, open brackets, commas, and colons. Her typographical arrangements are extremely open, too, producing a hybrid, audiovisual complex: pages are marked with horizontal and vertical repetitions of different sounds, and with clusters of differently sized fonts. Central to all such cacophony is the ongoing fact of the trees' silence but, as we move through the book, slowly this silence starts to acquire an ineluctable density. As Burk continues to ask questions, and as the trees continue to respond with non-semantic gestures, the poems form homes for what these gestures—these commas, open brackets, and colons—*might* mean. That silent "void" becomes, in the Nietzschean sense, productive: it is the catalyst for a reorientation of understanding. What the poems produce, in other words, is a field in which a conversation with trees becomes possible.

When we come to a poem like Murray's "Strangler Fig,"[40] we find that the virtual or speculative region mapped out by *Tree Talks* has suddenly, and gloriously, been made actual. "Strangler Fig" imbues its subject with an expansive syntax and fleshy vocabulary that is characteristic of much of *Translations*. Rather than a repressive, photorealist portrayal of the fig, or the framing of a static body, however, the poem is a moving form, accreting waves of flow. The point is not to ask of such poems "Is this real?" or "Could a tree *really* talk like this?" Rather, the point is to ask, "Is this *possible*?" In the case of "Strangler Fig," the answer is "Yes!" "Strangler Fig" constitutes the actualization of an imagined, affective world; the poem "particularizes the corporeality of figness," to quote John C. Ryan.[41] As it slowly drips down the page with viscous, fibrous weight ("muscling in molten stillness down | its spongy barrel"), fig and poem merge. The fig-poem's drive to completion is the realization of a translated mimesis: as the "bird-born" fig itself slowly takes hold of its host, as it compels everything beneath it into its own pronoun (the poem ends: "I one"), it also becomes the host for Murray's lexical prowess. Thus, the fig grows according to the skeletal structure of the poem, in the same way that a "regular" fig would use its host tree for support. Consequently, a continuous enjambed sentence slowly makes its way down the fourteen lines of a sonnet, a structure which the language simultaneously relies on and strangles into near disappearance: between each of the fourteen lines is a line break, all thirteen of which suggest the lines of an older sonnet form presiding like a ghost beneath this new articulation. The position of the nonhuman is not subsumed by the verse form, therefore; indeed, it could be quite the opposite (as fig subsumes host tree, so too does it subsume sonnet). Ryan, following Karen Barad, calls this "radically relational," where the voice of this vegetal form becomes inextricably bound up in the poiesis of the lyric.[42] Of course, the process at work here is an incredibly violent one, but if we are to attend seriously to the umwelt of the strangler fig, we cannot make moral judgments about such violence (at least, not if such judgments will prevent our engagement); the fig, and any poem attentive to it, can only exist via a most radical, suffocating form of relationality.

Interested in exploring such radical relationality, much of my creative practice of the last decade has involved many more-than-human collaborations. My previous poetry collection, for example, was a series of collaborations with particular places, where each place was as important to the process of composition as my own imagination.[43] Part of that collection's generative impetus came from an engagement with a broad Deleuzian sense of ethology: here, the world is alive with all kinds of entities, from rivers to bowerbirds to sand dunes, which "have distinctive ways of life, histories, and patterns of becoming and entanglement, [or] ways of affecting and being affected."[44] Broadly speaking, work like this is reliant on an expanded understanding of the different communities that compose a "culture," and the possibility of making creative translations across cultural and biological domains. As I turn to focus on specific creatures, bodies, or places, I am inspired by Sherry's contention that similarities of form can emerge at many different scales of existence,[45] and that by engaging with an organism I also am engaging with forces and flows—with what Deleuze and Félix Guattari would call "lines of flight"—which both constitute it and entangle it with others. Thus, for a while now, my poetry has not been "a monospecies event."[46] Through cultivating practices of ethological attention and creative translation, I have been committed to exploring the roles of nonhumans in composition.

My latest collection, *Lyre*,[47] could be based on the premise that nonhumans "are not some nicety or metaphorical convenience in poetry; rather, poetic intelligence is 'bound to animals' profoundly, and necessarily so."[48] Cultivating such attention has immediate and profound implications for the kind of poetry that one writes, and for the way we understand where that poetry comes from. Essentially, each poem in *Lyre* attempts to translate the expression of a particular Australasian species or landform (such as a house fly, a tea tree, or an estuary) into human language and, in order to do so, incorporates the discourses of biology and ethology, Indigenous knowledges, poetry from multiple languages, and my own poetry and detailed field notes. The collection implicates itself ecologically by working ethologically—that is, the collection establishes an interconnected community of different bodies by paying attention to the fleshy lives of particular bodies. But the actual "ecology" of species and landforms in the collection is based less on a scientific imperative to present an accurate cross section of any one ecosystem, and more on personal idiosyncrasies to do with where I live and travel, and what I am most interested in writing about. Rather than producing accurate surveys of a certain ecosystem, then, I wanted to engage with ecology as a *methodology* as well as an aesthetics, which means not only attending to ecological subjects in my writing, but also developing a compositional process that accounts for broader systems of human and nonhuman participants.[49] Literally, this involved combining material gathered from three distinct processes, often painstakingly. Drawing on a process that I learned from poet and scholar Jonathan Skinner, for each poem I assembled: (1) notes gathered from the relevant scientific, naturalist, and ethnographic literatures; (2) sections of relevant poetry and fiction from different languages; and (3) my own field notes, composed in particular sites in close proximity to the species or landform in question.

The first and second steps are largely scholarly, in that they require substantial research and reading, and synthesis of the information found therein. However, they also require keen editorial skills, as I am interested not only in the informational content of these sources, but also in the kinds of language they use. The challenge here is to mold together the various syntactical patterns and vocabularies of, say, a scientific article, a naturalist's blog, and a Brazilian poet, without erasing their particularities—in other words, to produce a blended text of multiple, divergent voices and rhetorical modes. My method for the third step is quite distinct from steps one and two. In order to attend to what can often be a disorderly profusion of signs produced by the species or landform in question, for the third step I engage in what Wendy Wheeler has called a "free-floating attentiveness."[50] I spend considerable time in the species' or landform's proximity, sketching out speculative translations of its language, and imagining, often quite wildly, the textures and rhythms of its umwelt. More theoretically, what I am doing in this process is trying to liberate some of those forces contained and produced by the organism or body enough so that I|we|language can enter into relation with them. The Deleuzian ethologist proceeds "by unravelling the reified solution posed by the organism" and untangling some of the affects that are clustered around and within it.[51] Of more interest than the animal as a body in itself—as an object that might be framed, for example—is what that animal (or plant, or river) is capable of doing affectively. When my focus turns to affect, it is trained "toward what escapes the nodal terms [of the organism]."[52] In other words, I am interested here in the transcorporeal potential of affective charge, of its capacity to strike out from one body to others, and to bind us together into a new becoming.

Of paramount concern has been that I do not simply echo the procedures of so many other poets, for whom animals are rarely more than objects to be looked at or to be described in representational language, as if they were little more than semiautonomous machines. I want to get closer (which often involves recognizing that I am actually quite far away), to write *toward* the other or, ideally, *with* it—recognizing, in other words, that "the other" is a region to which I am already bound. In more concrete terms, the biological roots of my humanity compose about 0.1% of my genome.[53] Already, I am almost entirely other, anyway. To quote Rose, "[h]umanity is an interspecies collaborative project."[54] But the desire to write toward must also be coupled with a concern for the outcome with relation to the subject in question. As Thom van Dooren, Eben Kirksey, and Ursula Münster point out, it is not enough just to celebrate "multispecies mingling." Rather, one must always ask, "who benefits when species meet?"[55] I am motivated by the challenge to "leave the mystery of the animal and its world intact while calling attention to its existence,"[56] and I have incorporated a variety of compositional procedures in order to do this. Most obviously, I resist adopting conventional and familiar modes of expression, any of which might induce an illusion that the subject of the poem is present in its totality, free of any kind of mediation or opacity. Furthermore, the texts often transgress the quadratic limits of the page, so that many poems feature lines that continue horizontally or diagonally across multiple pages, which can open multiple possibilities for the direction of reading|navigation. This means that there is never a single way of "reading" or understanding the subject of the poem: not only is the language many-sided, but often the eyes themselves cannot rest on a single, readily apparent path through the text.

In addition, using the list of materials above means that any perspective which claims a unique authority regarding the animal, plant, or landform always will be contrasted with, or contested by, other points of view. For example, in the composition of my field notes in step three, I might be inclined to venture well beyond the reality of my relation (or lack thereof) with a particular species, and enter highly speculative realms that have little basis in anything other than the whims of my imagination. However, the revision process ensures that the texts from such speculation must always contend with the different kinds of scientific knowledge, such as zoology and plant biology, gleaned from the first step. Necessarily grounded in collaborative, peer-reviewed inquiry, and in the repeated testing of hypotheses across different populations in multiple field sites, scientific literature is in a sense the direct antithesis to my own individual and highly localized speculations. By the same token, scientific literature is invariably absent of any account of interspecies interaction or entanglement, particularly concerning the researchers themselves with the species under analysis. By and large, the genre of writing in the natural sciences reflects what Rose and Plumwood would call a "hyperseparation" between the analytical mind of the scientist and an unthinking, "instinctive" natural world.[57] The use of my own field notes, as well as fragments from poems and fiction by other writers, goes some way to counteracting this separation.

On the one hand, this process is necessarily experimental, speculative, and improvisational. But in performing the experiment, I am affirming what Brian Massumi would call "an expressive force of variation" that, based on aesthetic and ontological orthodoxies, proceeds by unsettling them.[58] I am taking poetry into the Open, within and beyond which all kinds of unpredictable encounters might coalesce into radically unusual articulations of ecological fields. Famously, for Deleuze and Guattari, the improvisation might be a hazard , but "to improvise is to join with the World, or *meld* with it."[59] On the other hand, though, my interest in anthropomorphic form does not blind me to the perils of anthropo*centric* poetics: the extraordinary proliferation of stale, self-reflexive modes of confessional poetry across the world is confirmation that egocentric indulgence leads to little else but itself. I do not, therefore, want to leave things entirely to the predilections of my personal preoccupations. In order to cultivate what Anna Tsing has called the "art of noticing,"[60] I source different knowledge bases, each of which provides a detailed but distinct perspective on the creature, plant, or landform in question. Along with poetry, and occasionally fiction, in a number of languages, it is my hope that such knowledges can be combined into a "thick account" of the distinctive world of another body.[61] I want to emphasize, accentuate even, what Despret argues is the more-than-human condition of all art: the artwork is not the act of a single being, but is inscribed "in heterogeneous ecological networks."[62] For Despret, the very material of the work—here, my notes and quotes, not to mention my body and the bodies of those nonhuman subjects—is the source of captivation of both the audience *and* the artist: "the artist is not the cause of the work and … the work alone is not its own cause; the artist carries responsibility, the responsibility of one who hosts, who collects, who prepares, who explores the form of the work."[63]

These activities—hosting, collecting, preparing, and exploring form—are all of equal importance to my process. Indeed, when I began to explore other ways of articulating nonhuman language and expression, it became clear that such work would require not only an openness to different kinds of language, but also that the very forms of the poems reflect the vastly different umwelten of their subjects. Accordingly, the arrangement of the poems became as much to do with questions of design and visual composition as with those of vocabulary and prosody. In each case, the shape of the poem needed to reflect fundamental features about how the subject moves through the world and produces expression. But in changing the form of poetry so dramatically and frequently, the poems are not meant to offer some kind of simplistic mimesis of the subjects' shapes—something that a nature photograph could do far more effectively. Rather, I have wanted to let that "block of becoming," to use Deleuzo-Guattarian language again, produce something new entirely. In other words, the subject of the ethological poem needs to take part in a rewriting and reforming of the poem, to have an impact on how a poem can be read, and how it can look. In Evelyn Reilly's terms, this could be "a relational poetics," or a shift from the classical poetical and biological obsessions with naming the subjects under observation, to emphasizing "processes of interaction and change … to enact connections rather than to mark distinctions."[64] Here, somewhere in between grand, transcendent conceptions of "nature," "the environment," and complex ecological systems on the one hand, and the naming and objectification of bodies as if they were parts of a machine on the other, lie my ethological intentions; ever since Darwin, species are no longer "eternal 'forms' created by a rigid intelligence," but are "open, shapeshifting, creative events."[65] If the governing principle of one's process is a mutability prompted by unforeseen encounters with such events, then, following Andrew Schelling, the poetry that results "must also be emergent, open, shifty, organic, projective."[66]

As Ron Broglio argues, animals challenge those forms of language and representation that have become abstracted, disembodied, and regimented. "To make thought move and do real work at the horizon of the unthought," he writes, "representation should create a friction, reciprocity, and exchange between the human symbolic system of representing and the physical world shared with other creatures."[67] The point is to keep our epistemological loops open, to check our "stupendous capacity to ignore this very world within which our lives are sustained." It is because of such ignorance that the violence of the Anthropocene is perpetuated, its "extravagant wastage of lives and of earth."[68] A multispecies, collaborative poetics, therefore, implies a distribution of the body that thinks (or bodies that think), "creating a distribution of states or plural centres for valuing, selecting, and marking/making a world."[69] In formal terms, this implies a radical approach to the normative rules of grammar as they manifest in conventional syntax, to the relationship between sound and expression, and also to the relationship between meaning, marking, and language. Clearly, then, an aesthetics of realist representation, or a framing of the animal which wants to pretend that the frame does not exist, is inadequate to the task at hand—here, thought is taken nowhere, but rests instead with what it already assumes to be self-evident. Politically as much as aesthetically, the tendency to ensnare life within representations of value or identity must be avoided; what comes from the unknown cannot be packaged in predetermined forms. Rather, a new world needs to be envisioned, and for each encounter. The challenge posed by the ethological poem is analogous to the challenge posed by another animal, entangled with us even as it resists us.

Griffith University

Notes

1. Carrington, "Humanity".
2. World Wildlife Federation, "How Many Species?"
3. Rothenberg, *Thousand Mile Song*, 7.
4. Rigby, "Animal Calls," 117.
5. Baker, *Artist Animal*, 37.
6. Kirby, *Quantum Anthropologies*, ix–x.
7. Qtd. in Buchanan, *Onto-Ethologies*, 185.
8. Sherry, *Oops!*, 17.
9. Rose, "Cosmopolitics," 109.
10. Morton, "Ecologocentrism," 79.
11. Cooke, "Towards an Ethological Poetics." Whereas ethology generally refers to the observation of other animal species, I use the term to refer to the observation of all nonhuman (animal, plant, or otherwise) expression. Following Iovino, "every material formation, from bodies to their contexts of living, is 'telling', and therefore can be the object of a critical investigation aimed at discovering its stories, its material and discursive interplays, its place in a world filled with expressive forces." "Living Diffractions," 70.
12. Morton, *Ecological Thought*, 51.
13. Morton, *Ecology without Nature*, 99.
14. Morton, *Ecology without Nature*, 1.
15. Morton, "Ecologocentrism," 93.
16. Barad, qtd. in Haraway, *When Species Meet*, 289.
17. Balcombe, *Second Nature*, 29.
18. Despret, *What Would Animals Say?* 5; emphasis added.
19. Sherry, *Oops!*, 133.
20. Huggan and Tiffin, *Postcolonial Ecocriticism*, 16–17.
21. Rothenberg, *Why Birds Sing*, 54.
22. Garstang and Shepherd, *Songs of the Birds*; Rothenberg, *Why Birds Sing*, 47.

23. Rothenberg, *Why Birds Sing*, 49.
24. Gowdy, *The White Bone.*
25. Paull, *The Bees.*
26. Siccardi, *The Blackbird.*
27. Haraway, qtd. in Moe, *Zoopoetics*, 47.
28. Huggan and Tiffin, *Postcolonial Ecocriticism*, 17. For more discussion about the relationship of colonialist and nonhuman poetics, see Cooke, "The Songs of Others."
29. Heidegger, *Fundamental Concepts*, 177.
30. Plumwood, qtd. in Dooren, Kirksey, and Münster, "Multispecies Studies," 8.
31. Moe, *Zoopoetics*, 17.
32. Haraway, *When Species Meet*, 21.
33. Haraway, *When Species Meet*, 226.
34. Rose, "Slowly," 6–7.
35. Derrida and Wills, "Animal," 377.
36. Reilly, "Eco-Noise," 257.
37. Harrison, "On Composition".
38. Iijima, "Metamorphic Morphology," 288.
39. Burk, *Tree Talks*, 49.
40. Murray, *Translations*, 17.
41. Ryan, "Poetry as Plant Script," 132.
42. Ryan, "Poetry as Plant Script," 134.
43. Cooke, *Opera.*
44. Dooren, Kirksey, and Münster, "Multispecies Studies," 4.
45. Sherry, *Oops!*, 40.
46. Moe, *Zoopoetics*, 24.
47. Cooke, *Lyre.*
48. Shepard, qtd. in Moe, *Zoopoetics*, 28.
49. See Huggan and Tiffin, *Postcolonial Ecocriticism*, 13.
50. Wheeler qtd. in Baker, *Artist Animal*, 31.
51. Buchanan, *Onto-Ethologies*, 161.
52. Buchanan, *Onto-Ethologies*, 185.
53. Morton, *Ecological Thought*, 126.
54. Rose, *Wild Dog Dreaming.*
55. Dooren, Kirksey, and Münster, "Multispecies Studies," 16.
56. Broglio, *Surface Encounters*, xxxi.
57. Rose, *Wild Dog Dreaming.*
58. Massumi, *What Animals Teach Us*, 14.
59. Quoted in Baker, *Artist Animal*, 229; emphasis added.
60. Tsing, qtd. in Dooren, Kirksey, and Münster, "Multispecies Studies," 5.
61. Dooren, Kirksey, and Münster, "Multispecies Studies," 6.
62. Despret, *What Would Animals Say?* 5.
63. Despret, *What Would Animals Say?* 119–121.
64. Reilly, "Eco-Noise," 258.
65. Schelling, *Wild Form, Savage Grammar*, 77.
66. See Schelling, *Wild Form, Savage Grammar*, 78.
67. Broglio, *Surface Encounters*, xxxi.
68. Rose, *Wild Dog Dreaming*, 2.
69. Broglio, *Surface Encounters*, xxxi–xxxii.

Disclosure Statement

No potential conflict of interest was reported by the author.

ORCID

Stuart Cooke http://orcid.org/0000-0002-1688-642X

Works Cited

Baker, Steve. *Artist Animal*. Minneapolis, MN: U of Minnesota P, 2013.

Balcombe, Jonathan. *Second Nature: The Inner Lives of Animals*. New York, NY: Palgrave, 2011.

Broglio, Ron. *Surface Encounters: Thinking with Animals and Art*. Minneapolis, MN: U of Minnesota P, 2011.

Buchanan, Brett. *Onto-Ethologies: The Animal Environments of Uexküll, Heidegger, Merleau-Ponty, and Deleuze*. Albany, NY: State U of New York P, 2008.

Burk, Wendy. *Tree Talks: Southern Arizona*. Lockport, NY: Delete, 2016.

Carrington, Damian. "Humanity Has Wiped Out 60% of Animal Populations since 1970, Report Finds." *Guardian*, October 29, 2018. https://www.theguardian.com/environment/2018/oct/30/humanity-wiped-out-animals-since-1970-major-report-finds

Cooke, Stuart. *Lyre*. Crawley, NSW: U of Western Australia, 2019.

——. *Opera*. Parkville, VIC: Five Islands, 2016.

——. "The Songs of Others: Contemporary Poetics and the More-Than-Human." *Plumwood Mountain* 4, no. 1 (2017). https://plumwoodmountain.com/the-songs-of-others-contemporary-poetics-and-the-more-than-human/

——. "Towards an Ethological Poetics: The Transgression of Genre and the Poetry of the Albert's Lyrebird." *Environmental Humanities* 11, no. 2 (2019): 302–323.

Derrida, Jacques, and David Wills. "The Animal That Therefore I Am (More to Follow)." *Critical Inquiry* 28, no. 2 (2002): 369–418.

Despret, Vinciane. *What Would Animals Say If We Asked the Right Questions?* Translated by Brett Buchanan. Minneapolis, MN: U of Minnesota P, 2016.

Dooren, Thom van, Eben Kirksey, and Ursula Münster. "Multispecies Studies." *Environmental Humanities* 8, no. 1 (2016): 1–23.

Garstang, Walter, and James Shepherd. *Songs of the Birds*. London: John Lane, 1922.

Gowdy, Barbara. *The White Bone*. London & New York, NY: Picador, 2000.

Haraway, Donna J. *When Species Meet*. Minneapolis, MN: U of Minnesota P, 2009.

Harrison, Martin. "On Composition: Five Studies in the Philosophy of Writing." PhD diss., U of Technology, Sydney, 2010.

Heidegger, Martin. *The Fundamental Concepts of Metaphysics: World, Finitude, Solitude*. Bloomington, IN: Indiana UP, 1995.

Huggan, Graham, and Helen Tiffin. *Postcolonial Ecocriticism: Literature, Animals, Environment*. London: Routledge, 2010.

Iijima, Brenda. "Metamorphic Morphology (with Gushing Igneous Interlude) Meeting in Language: P as in Poetry, Poetry Rhetorical in Terms of Eco."

In *((Eco(Lang)(Uage(Reader))*, edited by Brenda Iijima, 275–292. Brooklyn, NY: Portable, 2010.

Iovino, Serenella. "The Living Diffractions of Matter and Text: Narrative Agency, Strategic Anthropomorphism, and How Interpretation Works." *Anglia* 133, no. 1 (2015): 69–86.

Kirby, Vicki. *Quantum Anthropologies: Life at Large*. Durham, NC: Duke UP, 2011.

Massumi, Brian. *What Animals Teach Us about Politics*. Durham, NC: Duke UP, 2014.

Moe, Aaron M. *Zoopoetics: Animals and the Making of Poetry*. Lanham, MD: Lexington, 2014.

Morton, Timothy. *Ecological Thought*. London: Harvard UP, 2010.

——. "Ecologocentrism: Unworking Animals." *SubStance* 37, no. 3 (2008): 73–96.

——. *Ecology without Nature: Rethinking Environmental Aesthetics*. Cambridge, MA: Harvard UP, 2009.

Murray, Les. *Translations from the Natural World*. New York, NY: Farrar, Straus and Giroux, 1992.

Paull, Laline. *The Bees*. New York, NY: Ecco, 2014.

Reilly, Evelyn. "Eco-Noise and the Flux of Lux." In *((Eco(Lang)(Uage(Reader))*, edited by Brenda Iijima, 255–274 . Brooklyn, NY: Portable, 2010.

Rigby, Kate. "Animal Calls." In *Divinanimality: Animal Theory, Creaturely Theology*, edited by Stephen D. Moore, 116–133. New York, NY: Fordham UP, 2014.

Rose, Deborah Bird. "Cosmopolitics: The Kiss of Life." *New Formations*, no. 76 (2012): 101–113.

——. "Slowly—Writing into the Anthropocene." *TEXT* 20 (2013): 1–14.

——. *Wild Dog Dreaming: Love and Extinction*. Charlottesville, VA: U of Virginia P, 2013.

Rothenberg, David. *Thousand Mile Song: Whale Music in a Sea of Sound*. New York, NY: Basic, 2008.

——. *Why Birds Sing: A Journey into the Mystery of Bird Song*. New York: Basic, 2006.

Ryan, John C. "Poetry as Plant Script: Interspecies Dialogue and Poetic Collaboration in the Northern Tablelands of New South Wales." *Transformations* 30 (2017): 127–149.

Schelling, Andrew. *Wild Form, Savage Grammar: Poetry, Ecology, Asia*. Albuquerque, NM: La Alameda, 2003.

Sherry, James. *Oops! Environmental Poetics*. Buffalo, NY: BlazeVOX, 2013.

World Wildlife Federation. "How Many Species Are We Losing?" Accessed November 11, 2019. http://wwf.panda.org/our_work/biodiversity/biodiversity/

Becoming D|other: Life as a Transmuting Device

By Astrid Joutseno

ABSTRACT

This essay explores the possibilities of extending the presence of an "I" of human (and nonhuman) self on the cusp of death or extinction. Reconfiguring loss of life as a story of hope through digital archiving, and viewing life as a device of transformation, the author weaves experiences and theory of illness time together with a new concept of D|other. The author employs D|other in illustrating what is taking place with material and digital relationality.

On Time

Living with stage IV breast cancer as a mother, doctoral student, and working artist places me on the verge of multiple edges. I hover on a threshold. This essay explores the potential of a digital archive becoming a hopeful resistance against annihilation by asking what if, instead of a finite ending, death were reconfigured as a leaving and life as a device of transformation? I portray archiving as an extension of subjectivity, a vehicle for the "I," which travels through life, narrative, and time. While focusing on these questions through a dying mother's perspective, I carry along a consideration for nonhuman lives on the cusp of a "double death": extinction.[1] In order to configure the "I" slipping between material and virtual existence, this essay introduces the concept of D|other.

This article begins with *illness time* as a moment from which to reconfigure subjectivity and life. I then introduce ways of archiving as an extension of mattering. Becoming D|other imagines a digital "I" with care and responsibility for others, starting from the point where M|other turns into D|other. At the end of the essay, I discuss death and extinction with the help of D|other, as well as the notion of leaving. As part of my intersectional feminist methods, I propose that questions function as careers of ideas, possibilities or pathways which lead to yet unknown places|times, where something remains to be shared. This method produces hope through a sharing of possible directions. As Aurora Levins Morales proposes, "questions can be as good as an answer. … When researching the history of lives whose history is poorly documented, often all we can do is point to the gap."[2]

My relationship to time is in flux. Time's unfolding is sensed differently from the perspective of a plant, a mineral, and a human in different phases of life. In Anna Tsing et al.'s *Arts of Living on a Damaged Planet*, the limitations of thinking of time as linear are illuminated through the example of lichens on gravestones: "Many kinds of time—of bacteria, fungi, algae, humans, and Western colonialism—meet on the gravestones of Petersham. The ghosts of multispecies landscapes disturb our conventional sense of time, where we measure and manage one thing leading to another. Lichens may be alive when we are gone. Lichens are ghosts that haunt us from the past, but they also peer at us from a future without us."[3]

As the presence of lichens suggests, time is entangled in the experiences of subjects. Karen Barad works with quantum physics, Afrofuturism, and Indigenous conceptions of time to break down the notion that time is a linear construction from material birth to death. She uses the concept of the *thick now* to illustrate the present as a stretched moment which contains the past and the future.[4] Becoming ill with what will be the cause of my time's end—an illness that will also, until that point, define my existence—has flooded my experience of time. I have landed in the present wanting to care for it. If living is layered with roles that define desires and mark wants, then living with the acute awareness of perishing touches time with something burning cold. The concept of *illness time* suggests that it is this deformation of time that interjects into imagining the loss of self—and then the creation of self with this knowledge. Illness transforms time, reshaping its meaning, deforming hopes, and illuminating the past anew. Illness questions time as a measurement of living. Illness relocates selfhood, queering the relationship to it and material bodies. Here I am thinking along the lines of *crip time* as put forward by Alison Kafer in *Feminist, Queer, Crip*. Kafer points out that crip time not only flexes time to adjust to disabled bodies and their differing needs against the social norms, but changes movement in time, thus criticizing time itself, arguing for the right to be disabled, different, and ill against the compulsory norms of able-bodiedness and the striving always to *get better*. Crip time works against curative time, which is the understanding of disability and illness as a passing state from which a future time will free us.[5]

What if the now of *illness time* could be accepted as a place of knowing and insight? What if the now of nonhuman measures of time were to be cast as a horizon of hope? The mutation that takes place on the cellular and DNA levels transforms a material experience of time, but that is not all. It metamorphoses relationships to environment, self, and other. It offers views into the vast perspectives of birds or clouds. I suggest that illness time is a *moment* from where people rearticulate perspectives to life, to the "I," to subjectivity, as well as to posthuman lives.

I acknowledge the limits of my ability to understand what constitutes living, belonging, existing, leaving traces—having an impact. When time as a concept crumbles, what appear are relations—everything in between—everything that binds, knots, or breaks away. Donna Haraway addresses this relationality with "tentacular thinking." Haraway extends her list of living tentacular animals to nets, networks, It critters, and clouds. From this perspective, life is lived "along lines … and not at points."[6] This description of intense relationality and Haraway's playful yet serious ways to combine the knowledges of the nonhuman and human guide me while I imagine life as a device of transformation. Attuned to troubles and without forced positivity, I follow Rosi Braidotti toward affirmative ethics, choosing to look for hope while acknowledging the sadness around losses in the Anthropocene.[7] Does this signify hope in the extinction of nonhuman lives or in the death of an ill person? No. Yet hope is necessary in bearing responsibility for the impact of living. In the following, I combine Braidotti's hope with Haraway's embrace of playful relationality to create a method of thinking about the interconnectedness of lives as networks, even when things appear at times unpoetic, accidental, or seemingly haphazard.

Archiving as an Extension of Mattering: Talking with My Child

One morning, I was reading the news on my phone as my child prepared for preschool. I casually told her that scientists are changing their interpretations of how the Neanderthal and Denisovan humans disappeared. Through examining small pieces of bone, researchers have concluded that Neanderthals, Denisovans, and Homo sapiens had children with each other more often than was believed. Their disappearance may have more to do with genetic meshing and gentle caring than with violent, intentional killing. I conclude that this is the most recent interpretation. My story sparks a discussion extending into death, life, transformation, and connections to artificial intelligence. After she leaves for preschool, I remain seated, scribbling in a notebook: oxygen binds me to this planet. From the perspective of the stars, I am bound here just as fish are to water. The environment keeps me alive. After death, there is no dependency on breath. I will ascend from the Blue, pass the thin oxygen of the upper stratosphere and go, unbound by the laws of gravity or the fragility of the body. Why do I hold on to this idea of the rise, which permeates religious imagery of death? Is the desire for artificial intelligence a secular dream of an afterlife or a desire to surpass vulnerability, a yearning to extend living and being into faraway places or different times? Does moving forward equal survival? As I narrated a story of change in time through loving "contamination" to my child, I added that her children may already be human only partly, with artificial intelligence for the other part. Her children will surpass the borders defining human and nonhuman. I sprinkle hope around. I frame the wish around the inventions of medicine as desire-to-overcome-vulnerability. (Affectively, I am tracing the path of curative time at this point.)

In our chat, we acknowledged the memories and life experiences of our great-grandparents—how they moved into the next generation through narratives and material objects that used to belong to them. We imagine continued subjectivity through machines after the human body no longer sustains life. We go further, pushing past the sadness of "what happens to bodies," the inevitable feelings of loss. Viewed this way, my vulnerability to the uncontrollable elements—the cells in my body—does not appear as much like a tragedy unfolding; instead, it becomes a process of transmutation. I am a joint. My child and I see that although my death will be the old-fashioned kind, we are on the cusp of extended life.

In telling the narrative of the past, I am doing my work as a M|other. I pass on a thread with which to weave the now and connect what is to come. I work with time. As I mourn the limits of my imagination and inability to know how this transition of selves will take place, I celebrate its possibility. I am filled with desire to remain beyond death. I yearn for connection to others. Connection is hope.

What about the extended subjectivities of nonhumans? Is their relational existence over without a digital archive or an artwork left behind? Is the idea of an archive as an extension merely limited to humans becoming trans, digital, or cyber? I feel guilt over what dies to facilitate my survival, but the evolution of affective responses may be a wasted language when it comes to the extinction of animals and plants. Mourning for extinction is dependent on human definitions of worth linked to capitalist ideas of everything as a resource, and the affects may be hindering my ability to see the lingering presence of what I think is gone. Tsing et al. remind me of the ghostlike ways in which the presence of what is gone is still affecting the planet: "Ghosts, too, are weeds that whisper tales of the many pasts and yet-to-comes that surround us. Considered through ghosts and weeds, worlds have ended many times before. Endings come with the death of a leaf, the death of a city, the death of a friendship, the death of small promises and small stories. The landscape grown from such endings is our disaster as well as our weedy hope."[8] Could the archive, then, be such a landscape growing out from endings into weedy hope?

Becoming Narrative

Media and technoscience are bubbling with images of the future seen through a techno-oriented late-capitalist lens, which posits development as an ideal that goes hand in hand with ownership and the resourcing of all that is living as data and capital, as Rosi Braidotti points out. Braidotti calls the focus on human DNA and gene technologies and digital extensions of life on social media a "matter-ialism."[9] She warns against technomysticism, saying that "projecting of the physical self into an artificial environment feeds into a dream of terminal identity outside the body, a sort of 'cybersubject' that feeds into the New Age fantasies of cosmic redemption through technology."[10] She underscores the futility of eternal undying subjectivity, warning against forsaking life over a fantasy of limitless existence in a material-less virtual space|time. I skirt around the danger, conjuring myself into a future—forsaking life yet going on as an "I"—both mourning the limited possibilities of existing in this body as a sick woman and imagining a material|virtual slide as resistance, which does not devalue the material and living, but redreams existence through a transmutation.

I am collecting an archive from which to burst out in ligaments of affect, thought, and expression after the material "I" has metamorphosed into another consistency. Currently, my work is to make sense of living while archiving the process. I work in multiple media simultaneously and view these strands as interlinked methods of becoming. For one, I have published an autobiography, *Viimeinen kirjani: kirjoituksia elämästä.*[11] Second, I am writing a doctoral dissertation at the University of Helsinki, weaving themes of mothering and illness into academic research. As a songwriter, I combine experience with poetics to explore life through music. It is through the writing and recording process of my album *From the Bed and Beyond* that I dealt with my first breast cancer diagnosis in 2014.[12] I narrated myself back into being in songs about the fear of death, sadness over what was lost, and the hope that continues to seep in. Life writing leaks into all areas of my work. As the location of "I" becomes readjusted by *illness time*, so does the political desire to keep working through this experience of living.

To theorize the "I" acknowledging the work done on digital life writing and automediation[13] and matricentric feminism[14] is to look for an understanding of multiple simultaneous forces: What if the "I" is a more resilient and lively being than material bodies? What if the green trunk in my living room full of handwritten journals from 1989 onwards is a version of "I" that has a more persistent life than my embodiment ever will? What if my digital diaries and writing stored in iCloud will outlive their writer by hundreds of years? What if "I" is ever present and the body a faulty and transient vessel (for want of a better word, although I dislike "vessel" as it reproduces the dichotomy between mind and matter)? What if ideas of strength and weakness smudge and hide the staying power of writing and narrating selves? What if the vulnerability of bodies makes them the sacred along with the planet, rather than dirty, shameful, and exploitable weaklings? And, as Braidotti asks, what if the "I" is just passing? What if effects instead of cognition are a fruitful way of approaching the world, as Braidotti formulates: the "I" is created and sustained "through affectivity and not cognition: as singularity, force, movement, through assemblages or webs of interconnections with all that lives. The subject is an autopoietic machine, fueled by targeted perceptions, and it functions as the echoing chamber of *zoe*. This nonanthropocentric view expresses both a profound love for Life as a cosmic force and the desire to depersonalize subjective life-and-death. This is just one life, not *my* life. The life in 'me' does not answer to my name: 'I' is just passing."[15] Yes, and what if the "I" in its passing through time and material embodiments is here to stay?

I am increasingly aware and intentional when I regard the archive that accumulates as I consciously exit life, while simultaneously accumulating traces, materials, digital fragments, and pieces of music. These are sensory gifts that work as threads, sewing me *here* after death has cleared away my being. I am interested in thinking through the archive as a nonhuman self—an archive of a life that relates to mine, but is not bound by material embodiment. This archive transmutes material|digital and human|nonhuman. Is there a border at which human becomes nonhuman? Is it a threshold between life and death or between material and digital?

Becoming D|other: From M|other to D|other

> Anthropocene landscapes of death and extinction are … inhabited by emergent and unexpected constellations of life, nonlife, and afterlife.[16]

What can it be that *lives* in the archive and is both digital and embodied? What about the trespasses I make through living into dying? Can there be voices or articulation in the unknowable? Can I push my imagination past the melancholy that comes with declaring death not as a threshold but as a finite cutting point? These questions slide from auto|bio into m|d|other.

D|other is an exercise in articulating what is taking place. Braidotti suggests that a posthuman affirmative ethics is constructed through imaginative thinking processes. She states that "transformative postsecular ethics takes on the future affirmatively, as the shared collective imagining that goes on becoming, to effect multiple modes of interaction with heterogenous others. Futurity is made of this."[17] I say that M|other is becoming D|other.

Adrianne Rich wrote *Of Woman Born* both to highlight mothering or being mothered as a unifying experience of all and to present arguments against patriarchal motherhood as an institution.[18] Since the 1970s, mothering has remained a troubling issue in feminist discourse. Becoming a mother continues to be the point at which equalities gained between gender divides become inaccessible to mothers.[19] This is the context in which I suggest M|other as a shift through which to address mothering and being a mother as an empowering, life-creating position resisting institutional definitions of gender or care. Where mother is a historical material position of repeated actions and care, M|other is entangled in the changes created through uses of technology and intersectional feminist thinking practices to subjectivity, reproduction, and data mining. M|other is a conceptual tying of matricentric feminism[20] and new materialist emphasis on intra-actions into imagining anew the roles of gender, care, and materiality. M|other is resistance to mothering as merely "women's work" or embodied, feminine, caring action. M|other underlines the material with the conceptual; it preserves the multiplicity of relational intersubjectivities in mothering; it acknowledges the other. M|other is becoming D|other—material mothering as care and life-giving, digital sharing, and, further still, living and creating as digital other, what has been called the cyborg or, as I will show, a D|other. By playing with this slippage and by starting with M|other, I involve the discourses and embodiments of mothers and the politics of mothering. I am thinking of being as living, existing, and being distinguishable as a subject. The other in this context is less a postmodernist concept and more a poetic expression of subjectivity defined as a technology of relating and belonging.[21]

D|other is not a concept wishing to end discussion, silence voices, or finalize debate. D|other arises like dry dust from the footprints of feminist thinking and thinkers. What foreshadows D|other and makes it possible to linger on the crossings between living and dying is what Barad calls intra-action in *Meeting the Universe Halfway*. In her words, "intra-action signifies the mutual constitution of entangled agencies." She claims that there are no preexisting separable agencies, but that distinct agencies "emerge through their intra-actions." Barad continues that "agencies are only distinct in relation to their mutual entanglement; they don't exist as individual elements."[22] When subjectivity emerges only through intra-action, it is time to think around subjectivity once again, to reconsider time and its relation to space again, as Barad suggests.[23] This emergent ontological shift calls for a leap into figuring out what might be possible and make sense in terms of posthuman ethics.

D|other emerges from the intersectional feminist insight of Audre Lorde, Aurora Levins Morales and Rosario Morales, Gloria Anzaldúa, and the theory and methods of intersectionality.[24] D|other is being imagined against the history of oppression that continues to be a reality in the present and affects lives at different intersections in distinct ways. With these grounding ideas, D|other leaps from material into digital to explore the issues of mothering and care, investigating the possibility of digital lightness, contingency, and multiplicity while holding the trouble that D brings: obsolescence.

D|other becomes meaningful through wordplay and storytelling, as portrayed by Haraway's "SF": "The tentacular ones tangle me in SF. Their many appendages make string figures; they entwine me in the poiesis—the making—of speculative fabulation, science fiction, science fact, speculative feminism, *soin de ficelle*, so far. ... SF is storytelling and fact telling; it is the patterning of possible worlds and possible times, material-semiotic worlds, gone, here, and yet to come."[25]

Haraway's "material-semiotic worlds" of SF reveal the importance of playfulness, willingness to investigate disorder, and the purposeful rearranging of thought in seeming randomness. I sketch D|other and M|other in this poetic construction:

In these names I hear the music of

mother   other
daughter   Other
material   other
digital
I see the separation between mother   M|Other   m|other
I see the differences between D|Other   Daughter   Other

D   death
M   memory

I am playing with transition
sensing the weight of history with Other
Considering a possibility for posthuman m|d politics
throwing out what is on the edges, the peripheries of what I can consciously fathom
making notes, first drafts, leaving traces, sharing,
taking this down now so the sketch is here for anyone to go on
So that when my last breath unites with oxygen and I ask for no more

> there is something here to connect with …
> Or at least cartographies of thoughts and hopefulness to trace through while looking maybe, for the narratives of "I" or possibly theories for transmutations.
>
> This is the poetics of seeing what is connected: the mother to the care the M|other to love manifested in life-giving but not to gender as a binary or a determiner of who should bear responsibility for this coexistence. The other to the separation of experiencing selfhood …

D|other came to me while formulating a way of conceptualizing what takes place with mothering on mommy blogs. D|other helps in picturing the entanglement of posthuman, nonhuman, gender, self, subjectivity, identity, materiality, digitality, and life. It is a folded concept of becoming, to echo Sara Ahmed's investment in the plurality of feminist concepts.[26] I am putting this in language, but you are with me as D|other. Importantly, so are the thinkers and doers of the past and this moment. And so is nonhuman life, so is what is to come, even though it has not been born, and while it exists simultaneously and does not follow linearly from the present.[27] D|other is subjectivity and activity which, through M|other, channels the wisdom of caring for another. It is related to Braidotti's affirmative ethics: "the desire to endure in time and thus clash … with the deadly spin of the present" and the understanding that freedom can be extracted "through the understanding of our bondage."[28] It is an understanding in actions of the idea that there is no distinction between self and other or material and digital life. D|other is about assembling a subjectivity that is distinguished from others and relates to them in movement. D|other lives virtually; she lives in connections and ties made real through a multiplicity of condensations. She lives in archives of work, fragments of writing, affects, impacts, and code. As far as D|other is a self, she spans material life expectation, transmuting the meaning of death in its wake.

In addition to expressing deep relationality, entangled commitment to caring for others, D|other yearns for extension to go on, to relate, to matter, to be matter, to breathe. Thinking through evaporated embodiments, digitalized, reorganized positions of being, there to grasp what remains is D|other. In the case that oxygen-lung-living becomes impossible, I am following the desire to restate loss as transition. I remain hopeful for an interpretation of an ending as fluid transformation, which annihilates the possibility of remaining in familiar form. I mourn the loss of materiality and try to remove all unknowable effects of the loss preemptively. Still, I realize the futility of such a task. Underlying this poet(h)ics of D|other are connections that flow into each other, extending notions of living and mattering.[29]

Survival Is Not a Theory[30]

Material bodies that do not conform to a norm become visibly troubling. Brownness and blackness continue to trouble white institutions. Disabled bodies wreak havoc with ableist structures. Academic systems and political and social orders continue to churn out normative whiteness and heterosexuality. Institutionalized whiteness requires that those not inhabiting norms become visible—"sealed by whiteness," as Ahmed describes the dynamic.[31] Ahmed also points out that "those who do not quite inhabit the norms of the institution are often … given the task of transforming these norms."[32] I write about D|other not as another colorless, bodiless contortion of a white hallucination, but as a colorful, wide-eyed, spilled-open, and leaky concept for something on the move. D|other is about reaching toward, connecting, and sharing, not so that those already marginalized share, give, and justify their positions, but so that entangled politics of living (in time, location, and human as well as nonhuman forms of living) propel those with privilege toward action and empower those without it. Lorde wrote in the 1980s about the need to move conceptually toward a more unified but nuanced understanding of difference: "Can any one of us here still afford to believe that efforts to reclaim the future can be private or individual? Can anyone here still afford to believe that the pursuit of liberation can be the sole and particular province of any one particular race, or sex, or age, or religion, or sexuality, or class? Revolution is not a one-time event. It is becoming always vigilant for the smallest opportunity to make a genuine change in established, outgrown responses; for instance, it is learning to address each other's difference with respect."[33]

It is frustrating that despite changes in the constitution of everyday communication practices (in the northern hemisphere where the Internet is most prevalent) and the overbearing advancement of late capitalism, the strides made toward Lorde's vision have been elliptic and slow (also disrupted and obscured, particularly in terms of the global digital divide). Digital identities do not make discrimination and racism disappear. Transmutations between material and digital depend on the material embodiment. The digital "I" is inherently entwined to it, as Lisa Nakamura illustrates in *Cybertypes* in her discussion on cyber identities in films.[34] D|other is no quick fix. My aim is that it becomes a practice, a tool with which to hold the particularity of life in color, resisting and against norms, while honing in on a movement or a transformation, or a wiring through which to breech the disturbed fracturing of memory boards and neural paths, the separation of the living based on skin colors and power distribution or class, but also the kinds of organisms and arrangements that other lives and matter have. Posthumanist thought, with its language of instability and fragmentation, combined with late capitalism is instigating dangerous articulations of insecurity which drive those with power toward separation. Identity politics becomes a game over who deserves to live. This is the creation of extinction, genocide, war, murder, violence, death, and silence. This is the regurgitation of arguments for white power or men's superiority over women, and the persistent misconception of trans identities as dress-up games and the neurotic fight to distinguish between living matter that is human and nonhuman. This is the compartmentalized map of "living things" which does not allow for the memories of those lost, the recognition of ghosts and those who have no voice, or those who had a voice but who suffered erasure.

One aspect of D|other is thinking about the role of archives, both in digital formats and otherwise. In 2017, bell hooks placed her archives at Berea College, Kentucky, in The bell hooks Institute.[35] She chose this home for her life's work (writing, papers, talks, and acquired works of art, but also artifacts) to make sure that it does not disappear at her life's end. She is acting against the historical fact that black women's bodies of work slip into obscurity quickly. hooks is imagining ahead, thinking as a D|other with respect and love for her value and contribution in life, extending its reach beyond her personal gain or control. This is a key aspect of D|other—an ethics of care extending beyond the "I," rather than the wish to have power to control an outcome.

Archives of women artists and writers have a history of disappearing. Even those who have had moderate to big success in their lives have become a target of erasure after their material life has come to an end. These disappearances take the form of publishers who decide to press a title no longer, institutions such as universities that forget to mention a researcher's contribution or stop teaching her in their syllabus, record labels that stop rereleases, and art institutions that forget to include works in exhibitions. They may also be the result of an email account becoming inaccessible or a digital archive disappearing. The disappearances go unnoticed and the narrative gives in: suddenly it is unclear why a woman's work mattered. *Who was this mother of three? Who was this brown transwoman? What was the context of her life's work?* Women's archival disappearances constitute relatives who do not value their loved ones' legacy or feel shame, jealousy, or ambiguity toward their perspectives. Documents get lost, misplaced, or badly organized, or maybe they have never been ordered but made sense to the living creator|artist, yet now appear a mess with no clues to what relates to what.

Rather than a conscious malicious erasure, it is an accumulation of non-actions that constitutes extinction and leads to voicelessness and lack of traceable genealogies. Scientists such as botanists and zoologists have used archives as survival strategies for extinct plants and animals, the nonhuman lives unable to download themselves into a digital memory or arrange for a library to house their story. But the archival survival of nonhumans is mediated by humans, thus becoming survival through human affects. I reside with contradictions. I grieve the losses of lives, voices, and narratives. Who let go? Who could afford this loss? Who took their place? I cry over my sense of limited time.

Extinction Is Not a Metaphor

Leaving

It has been argued that death signifies the end of agency and the possibility of communication, as there have been no messages from those who died. Is this so? What constitutes a message? Does it necessitate an intention from the sender? Does it matter when the message was stuffed in the bottle? What is the role of the delivery method? Can "I" not exist if my impact and voice are still carried? What about the effects that the lived past continues to have—the ghost winds that keep haunting the present, as Tsing et al. name the presence of nonhuman and human lives past?[36] I am thinking of lives that can still be experienced, remembered, thought, felt, and heard after death, often in surprising ways. What if a nonsecular understanding of dying was reconsidered as leaving? *What if the D is for death?*

Leaving has its own poetry. It is not as final sounding as extinction and death. Leaving suggests volition. It contains agency. I imagine death as a leaving, after which an archive becomes a presence. It becomes a life in the sense that it extends the time of the subjectivity entwined with others. It is another articulation of being D|other—being *here* and *doing time* in more ways than one. It is an active exit, constructing an entry. D|other becomes a process of dying and becoming, a transmutation, from M|other to D|other.

Not only do archives produce existence as D|other, but grief also conjures up the existence of what is already gone. Here I use "conjuring" not as a term suggesting a magical level of metaphysics, but to point out that, as part of this discourse, D|other invites the language of magic to reflect on lived life. Then, in a visceral manner, the goneness of what has left becomes questionable.

In this way of seeing, nothing appears as it used to. Life is no longer the value itself, the late-capitalist currency of choice. Life becomes a device through which "I" transforms and takes new shape in both material and virtual forms. Humans and nonhumans pass through life. Life becomes a vehicle, the vessel with which existences intensify and die down. Grief begins to transform. Attachments appear differently. The elusive "I," like a diva on a corner stage of a smoky bar, becomes the center of attention.

Extinction

> As a creative writer, I know that the story of something broken becomes something whole.[37]

Residing in this M|D slippage is about locating narratives of what living means in the Anthropocene. It is a landscape where distinguishing between human-made disaster and natural disaster is exceedingly difficult.[38] The blurring of lines between selves and others is messy, while past and future times haunt the present. What needs to be extended are capacities of imagining, dreaming and hoping with intention, and looking back, or discussing the affects curling around exits and extinctions—embracing death and illness time as a painful but vital potency of transformation through resistance and feeling because what reaches over time and space creates openings for unexpected yet unimagined (but imaginable) others. This creates possibilities for ethical action.

D|other rises from the ashes of burnt thought and the inflamed relationality of what once appeared stable. It is born from lost life, material and virtual. Extinction is not an equalizer, as death at this point still is. We are all going to die, but not all of us are going to be extinct. Nils Bubandt describes what is at stake: "In the Anthropocene, life is already geologic. In this geological ghost vision, the present proceeds from the future, because the possibility of co-species survival depends crucially on what we humans are going to do now, in the midst of an increasingly given fate of ruination and extinction."[39] This perspective toward the present moment produces a need to act.

Extinction affects peoples, organisms, and species differently. My mourning the nearing end of a personal life is not comparable to the experiences and threat of extinction which continue to annihilate the life of thousands of species and indigenous peoples, or the threat to and killings of black, brown, trans, and queer people. Again, the question of human-made disaster versus natural biological failure reaches me like an echo. Yet, living with an incurable illness drives me toward acute awareness of the slippery edges of societal norms, and of dying as a part of living. The cost of avoiding this awareness in Western culture at large manifests in mass killings and the unethical, uncaring treatment of the living as an exploitable resource.

One purpose of grief and sadness is to agitate toward change. Remembering is key. Morales underscores the role of memory, telling stories of fiction, and imagining the future: "Our job as people with memory, as participants in this moment in time is to keep looking for and examining the implications of the stories we are told."[40] Her words highlight the role of life narratives, of telling about pain, and the importance of listening, archiving, and reading. Braidotti suggests that much of contemporary (advanced-capitalist) discussion and feminist discourse around pain is stuck in a mode of claims and compensations which glorifies suffering. This means that "pain within a discourse and social practice of suffering … requires rightful compensation." With her creation of affirmative ethics, Braidotti swims upstream to discover alternatives to the discourse of compensation. She suggests that "ethics consists in accepting the impossibility of adequate compensation—and living with the open wound."[41] It is my understanding that this living with the wound produces narratives of life. Braidotti imagines an ethics which respects pain and acknowledges it as real and hurtful as well as unjust, but which instead of compensation and retribution seeks hope: "Accepting the impossibility of mutual recognition and replacing it with mutual specification and mutual codependence is what is at stake in postsecular affirmative ethics. The ethical process of transforming negative into positive passions introduces time and motion into the frozen enclosure of seething pain. It is a postsecularist gesture of affirmation of hope, in the sense of creating the conditions for endurance and hence for a sustainable future."[42] I add that to visualize a life without "me," a time post "I," is part of the beginning of an ethics of sustainability, which I imagine as a route elsewhere from violations (if not healing, because it is not always the goal) or acceptance of the story of the "I" with its fragility.[43] And which, from this perspective, produces grounds for earthly sustainability—that is, the continuity of life in ever-transforming shapes. Considering life as the vehicle for subjectivity, for the "I" in passing, helps to dislodge this emergent ethics from the blame|compensation discourse and the unsettling grief over death through focusing on life as an instrument of this transformation. With hope located in intentional imagination, I write in the present time, holding both grief and joy, feeling connected to past and future times and places.

University of Helsinki

Notes

1. Rose, "Shimmer."
2. Morales, "Aurora Levins Morales."
3. Tsing et al., *Arts of Living*, 10–11.

4. Barad, *Meeting the Universe Halfway*; "Karen Barad."
5. Kafer, *Feminist, Queer, Crip*, 28–29.
6. Haraway, "Tentacular Thinking," 32.
7. Braidotti, "Politics."
8. Tsing et al., *Arts of Living*, 6.
9. Braidotti, "Politics," 202.
10. Braidotti, "Politics," 202–203.
11. Swan, *Viimeinen kirjani.*
12. *From the Bed and Beyond* was released on the Soliti label in 2017.
13. See Poletti and Rak, *Identity Technologies.*
14. See O'Reilly, *Matricentric Feminism.*
15. Braidotti, "Politics," 210.
16. Bubandt, "Haunted Geologies," 137.
17. Braidotti, "Politics," 216.
18. See Rich, *Of Woman Born.*
19. O'Reilly, *Matricentric Feminism*, 2.
20. O'Reilly, *Matricentric Feminism.*
21. See, for example, Smith and Watson, *Reading Autobiography* and Poletti and Rak, *Identity Technologies.*
22. Barad, *Meeting the Universe Halfway*, 33.
23. Barad, *Meeting the Universe Halfway*, 33–34.
24. Lorde, *Your Silence*; Morales and Morales, *Getting Home Alive*; Anzaldúa, *Borderlands/La Frontera*; Lykke, *Feminist Studies.*
25. Haraway, "Tentacular Thinking," 31.
26. Ahmed, *Living a Feminist Life.*
27. Barad, *Meeting the Universe Halfway* and Braidotti, "Politics."
28. Braidotti, "Politics," 215.
29. Further along the way (but not in this article), with this M|D slippage, I wish to account for and create language for witnessing and experiencing such extensions of life and living that transform what constituted previous definitions of life and death, human and nonhuman, or posthuman.
30. See Lorde, *Your Silence.*
31. Ahmed, *Living a Feminist Life*, 133.
32. Ahmed, *Living a Feminist Life*, 135.
33. Lorde, *Your Silence*, 127.
34. Nakamura, *Cybertypes*, 43–54.
35. See http://www.bellhooksinstitute.com
36. Tsing et al., *Arts of Living*, 1–2.
37. Morales, "Aurora Levins Morales."
38. Bubandt, "Haunted Geologies," 126–136.
39. Bubandt, "Haunted Geologies," 136.
40. Morales, "Aurora Levins Morales."
41. Braidotti, "Politics," 212–213.
42. Braidotti, "Politics," 214.
43. Braidotti, "Politics," 212.

Disclosure Statement

No potential conflict of interest was reported by the author.

ORCID

Astrid Joutseno http://orcid.org/0000-0003-1942-8907

Works Cited

Ahmed, Sara. *Living a Feminist Life*. Durham, NC: Duke UP, 2017.
Anzaldúa, Gloria. *Borderlands/La Frontera: The New Mestiza*. San Francisco, CA: Aunt Lute, 1987.
Barad, Karen. "Karen Barad: Troubling Time/s, Undoing the Future." Faculty of Arts, Aarhus U, December 8, 2016. https://www.youtube.com/watch?v=dBnOJioYNHU
——. *Meeting the Universe Halfway: Quantum Physics and the Entanglement of Matter and Meaning*. Durham, NC: Duke UP, 2007.
Braidotti, Rosi. "The Politics of 'Life Itself' and New Ways of Dying." In *New Materialisms: Ontology, Agency, and Politics*, edited by Diane Coole and Samantha Frost, 201–218. Durham, NC: Duke UP, 2010.
Bubandt, Nils. "Haunted Geologies: Spirits, Stones, and the Necropolitics of the Antropocene" In *Arts of Living on a Damanged Planet*, edited by Anna Lowenhaupt Tsing, Heather Anne Swanson, Elaine Gan, and Nils Bubandt, 121–141. Minneapolis, MN: U of Minnesota P, 2017.
Haraway, Donna. "Tentacular Thinking." In *Staying with the Trouble: Anthropocene, Capitalocene, Chthulucene*, 30–57. Durham, NC: Duke UP, 2016.
Kafer, Alison. *Feminist, Queer, Crip*. Bloomington, IN: Indiana UP, 2013.
Lorde, Audre. *Your Silence Will Not Protect You*. London: Silver, 2017.
Lykke, Nina. *Feminist Studies: A Guide to Intersectional Theory, Methodology and Writing*. London: Routledge, 2010.
Morales, Aurora Levins. "Aurora Levins Morales: 'Memory Is Our Soil: Bringing History into the Commons.'" UMassHistory, U of Massachusetts, April 26, 2018. https://www.youtube.com/watch?v=2MOBA7AUl5Q
Morales, Aurora Levins, and Rosario Morales. *Getting Home Alive*. Ithaca, NY: Firebrand, 2013.
Nakamura, Lisa. *Cybertypes: Race, Ethnicity, and Identity on the Internet*. New York, NY: Routledge, 2013.
O'Reilly, Andrea. *Matricentric Feminism: Theory, Activism, and Practice*. Bradford, ON: Demeter, 2016.
Poletti, Anna, and Julie Rak. *Identity Technologies: Constructing the Self Online*. Madison, WI: U of Wisconsin P, 2014.
Rich, Adrianne. *Of Woman Born: Motherhood as Experience and Institution*. New York, NY: Norton & Company, 1995 [1986].
Rose, Deborah Bird. "Shimmer: When All You Love is Being Trashed." In *Arts of Living on a Damaged Planet*, edited by Anna Lowenhaupt Tsing, Heather Anne Swanson, Elaine Gan, and Nils Bubandt, 51–64. Minneapolis, MN: U of Minnesota P, 2017.
Smith, Sidonie, and Julia Watson. *Reading Autobiography: A Guide for Interpreting Life Narratives*. 2nd ed. Minneapolis, MN: U of Minnesota P, 2010.
Swan, Astrid. *Viimeinen kirjani: kirjoituksia elämästä*. Helsinki: Otava, 2019.
Tsing, Anna Lowenhaupt, Heather Anne Swanson, Elaine Gan, and Nils Bubandt, eds. *Arts of Living on a Damaged Planet: Ghosts and Monsters of the Anthropocene*. Minneapolis, MN: U of Minnesota P, 2017.

Writing the Lives of Plants: Phytography and the Botanical Imagination

By John Charles Ryan

ABSTRACT

Phytography refers to human writings about plant lives as well as plant writings about their own lives. The author conceptualizes phytography in terms of vegetal intelligence, behavior, corporeality, and temporality. Narrating the complex worlds of plants, phytography uses a variety of formal strategies to advocate new possibilities for human-flora relations.

> There is a wild apple on Nawshawtuct Hill in my town which has to me a peculiarly pleasant bitter tang, not perceived till it is three-quarters tasted. It remains on the tongue. As you eat it, it smells exactly like a squash-bug. It is a sort of triumph to eat and relish it.
>
> — Henry David Thoreau, *Wild Fruits: Thoreau's Rediscovered Last Manuscript.*

Introduction

Recent popular botanical nonfiction nourishes a long-standing public fascination with the mysterious inner worlds of plants.[1] Weaving between scientific exposition and narrative reflection, forester Peter Wohlleben's *The Hidden Life of Trees*, published in English in 2016, appraises the occurrence of friendship, language, love, and communication in the arboreal domain.[2] Organized around recurring visits to twelve tree personae, moreover, biologist David George Haskell's *The Songs of Trees* foregrounds the complexities of sonic expression within forest communities. Framing vegetal cognition as intrinsically networked, Haskell claims that "part of a plant's intelligence exists not inside the body but in relationship with other species."[3] Indeed, the appearance of these and other examples of botanical nonfiction parallels the development—particularly over the last fifteen years—of the field of plant cognition and behavior.[4] An upshot of this increasing interchange between botanical science and narrative is the unabashed characterization of plants as "intelligent" by Haskell, Richard Mabey, Wohlleben, and other nonfiction writer-naturalists. Nevertheless, interlinked generalist and technical interest in the percipience of vegetal life—beyond its instrumentalization as food, fiber, and medicine—is nothing new. An infamous case in point is *The Secret Life of Plants*, published in 1973 by journalists Peter Tompkins and Christopher Bird. Partly an elaboration of biophysicist Jagadish Chandra Bose's experimental work in the early twentieth century, the popular though controversial account—deemed spurious and esoteric by many scientists—later became a documentary featuring a soundtrack by Stevie Wonder.[5]

For Bose, the possibility that plants signify their lives through forms of language was more than metaphorical. In an address delivered in 1911, he postulated that vegetal "script"—disclosed through technological innovations—would reveal the interior worlds of plants that otherwise would remain concealed from human awareness.[6] To be sure, Bose's scientific writings intersect with some contemporary botanical nonfiction through a shared belief that elusive vegetal lives can be unraveled and, thus, rendered transparent. The development of instrumentation—for instance, Bose's crescographs—and the enactment of reductive modes of thinking about plants constitute the means to decipher vegetal being-in-the-world. A prominent aspect of the marginalization of individual botanical lives in these nonfiction accounts and others is the collectivization of plants' percipient faculties. In *The Songs of Trees*, for instance, the individuation of the balsam fir serves as a framing device for the larger scientific narrative of the species and its ecological relations.[7] Toward a view of plants as networked memes with transcorporealized intelligence, Haskell even summons Virginia Woolf's assertion that real life is "the common life, not the 'little separate lives which we live as individuals'" and, moreover, that "common life is the only life."[8] In response to Haskell's assertion, in this article, I ask the following: Where are the singular lives of plants in botanical prose and poetry? The problem, as I understand it, is that Haskell and other plant writers, at times, cannot see the trees for the forest. My intention here is neither to trace nor deconstruct this problem philosophically, as plant-thinker Michael Marder has already done,[9] but instead to delineate *phytography* as the writing of plants—as our writing about their lives and their writing about themselves and, possibly, about us and us in relation to them.

In conceptualizing phytography in terms of posthumanist life writing and proposing its main tenets, I examine nonfiction and poetry that I believe counter a tendency in contemporary botanical writing to privilege plant communities and networked intelligence over individually percipient—and sentient—personae.[10] Cultivating careful and strategic forms of anthropomorphism,[11] these phytophilic writers—from Henry David Thoreau in the nineteenth century to poet Wendy Burk in recent years—echo Marder's assertion that "the plant is at once the most singular and the most general being."[12] Their work offers insights into the complex lives of trees, shrubs, and herbs through various focal points. These include particularization (attention to individual plant characters), percipience (plants as intelligent, responsive, and agentic beings), corporeality (plants as embodied individuals located in time and space), temporality and seasonality (the changeability of flora over time and seasons), emplacement (the influence of place on plantness and vice versa), language and signification (the interpellations, interpolations, and communicative modalities specific to vegetal life), historicity (the intertwining of botanical and human histories), and mortality (the decline and demise of plants as meaningful events prompting human mourning, memorialization, and elegy). Within the overarching phytographic frame, I put forward two posthumanist life-writing principles relevant to the dialectic of auto|biography: *writing-with* and *writing-back*. The first term denotes more-than-human life writing composed in dialogue with living plants, whereas the second signifies the ways in which plants write their own lives—sensorially and materially—irrespective of human mediation. The article then concludes with an overview of my ongoing experiment in field-based poetic composition, "Gorge: Scriptorium | Tree | Excubitorium," carried out in conjunction with the lively flora of the Northern Tablelands of New South Wales, Australia. In "Gorge," collaborative human-botanical script hinges on *writing-with* plants while allowing their *writing-back*—their own expressions, utterances, gesticulations—to suffuse the poetic text.

Life Writing and the More-Than-Human: Theoretical Perspectives

Writing botanical lives involves negotiating various compositional hurdles originating in commonplace discourses about, and perceptions of, plants. Of course, in contrast to humans, mammals, and birds, vegetal life is mostly sessile and, therefore, difficult to perceive on an everyday basis as doing anything noteworthy, except when flowering, fruiting, or dying. Traditionally regarded as mute and passive—"they hardly move and make no noise"—plants adhere to temporalities sharply divergent from mobile creatures.[13] They orchestrate evolutionary processes, such as photosynthesis, essential to the biosphere yet grossly underappreciated by humankind.[14] Machinic and aesthetic tropes, moreover, dominate ways of thinking about plants, reducing their inner workings to "control circuits" and likening their outer forms—leaves, trunks, roots—to "beautiful objects."[15] Confronted by the formidable otherness of vegetal beings, we attempt to exert linguistic control over them through taxonomic designators (species, genera, families, varieties), sexually-fixated metonymies (flowers, blossoms, blooms, fruits), and reductive scientific terminologies (specimens, samples, compounds, active ingredients). What is more, a prevailing instrumentalization characterizes most human relations to flora, formalized in the paradigm of economic botany, or the "biology, culture and utilisation of plants and plant products."[16]

The prevalence of utilitarianism, instrumentalization, and reductionism in our dealings with plants constrains the emergence of more-than-human life writing attuned to their intelligence, sentience, and other complexities, as demonstrated to an increasing extent by scientific research.[17] Notwithstanding their opaqueness—at least from a human standpoint—individual plants should be regarded as meaningful narrative subjects in their own right. My contention here reflects the most inclusive and biocentric sense of *life writing* as "writing that takes a life, one's own or another's [including an animal's or plant's], as its subject. Such writing can be biographical, novelistic, historical, or explicitly self-referential and therefore autobiographical."[18] Instead of narrowing plants to their uses and appearances—and so negating their diverse capacities—phytography engages "the botanical imagination" as the relational, intercorporeal, and dialogical opening of narratives to vegetal being.[19] As we envision plants as auto|biographical subjects, so they imagine us back in an interplay of imaginings; as we write the lives of plants, so they write their own lives—and ours. Plants write-back into auto|phytographical accounts as their unique articulations weave into the fabric of diverse shared narratives.[20] This posthumanist view shifts phytographical writing from concerns of textual representation toward the interbraiding of the authorial (human) self with the creative agencies of vegetal lives.[21] Writing-with plants necessitates openness to their material-sensorial significations—their feedback, edits, and criticisms, if you will—which trouble the assignation of language to the human domain.[22]

From a phytographical perspective, "to show interdependency, human language would be replaced by diverse communications and interactions."[23] The challenge for the human writer of plant lives is to sense, listen, and allow in a state of suspended judgement comparable, for instance, to Edmund Husserl's idea of phenomenological bracketing.[24] Writing-with plants—and in dialogue with their writing-back—encourages a movement of human awareness of flora between a preoccupation with *botanical life* (homogenizing, based in the taxonomic ordering of nature) and an embracing of *botanical lives* (heterogenizing, grounded in an empathic regard for individual plants as subjects with particular lifeworld experiences). Such a transformation in outlook, I believe, lays the groundwork for a posthumanist theory and practice of phytography. In this regard, literary scholar Nancy K. Miller argues that "in autobiography [and other forms of life writing] the relational is not optional. Autobiography's story is about the web of entanglement in which we find ourselves, one that we sometimes choose."[25] Although not explicitly referring to the more-than-human, Miller urges the reconfiguration of humanist tenets of life writing through investigation of entangled subjectivities in narratives. In coming to theorize posthumanist life writing about animals, however, scholars have largely overlooked plants. Cynthia Huff and Joel Haefner, as an example, theorize the representation of non-human animals in "animalographies" or "narratives 'by' and about animals, in which human beings control what animals say, hiding behind an image and an inscription humans create."[26] According to Huff and Haefner, these kinds of zoo-centric stories constitute anthropomorphizations as "humans reconstruct their own subjectivities in the voices of animals."[27] In animalographies, the human writer "ventriloquizes the animal's voice allegedly to tell his story."[28] Nevertheless, some works, such as the children's novel *A Dog's Life: The Autobiography of a Stray* by Ann M. Martin, enable young readers to gain an appreciation of the difference between ethical and unethical relations to animals.[29]

To be certain, life-writing scholars have also theorized biography and autobiography in relation to poetry. Terms such as *biographical poetry*, *biographical verse*, and *verse biography* denote "poetic texts that tell the life story of another [human or more-than-human]."[30] Biographical poetry often discloses the author's perception of a character's ethos (moral orientation) and dianoia (the rationality of his, her, or, in the case of a plant, its thinking) while at the same time generating cultural and historical insights.[31] In contrast, autobiographical poetry "denotes a clear referential connection between the drama of the work and the drama of [the poet's] own life."[32] Some poems—especially in the confessional tradition of W. D. Snodgrass, Sylvia Plath, and others—are unequivocally autobiographical narratives where the poem's speaker *is* the writer.[33] In certain cases, however, the broader aesthetics and polemics of confessional poetry decouple it from pure autobiography.[34]

Phytography, then, is critical posthumanist life writing about more-than-humans that pivots on the potential of collaborating and coauthoring narratives with plants, notwithstanding their profound otherness. The phytographical genre emphasizes intersubjectivity, interactivity, and relationality, resisting "unimodal" approaches predicated on strongly humanistic tenets.[35] Although entailing a critique of science—and, specifically, reductive taxonomic, morphological, biochemical, and biogeographical approaches to plants—the move toward phytography does not constitute a wholesale rejection of botany. Instead, contemporary phytography involves dialogue with emerging forms of botanical research—namely, plant signaling, cognition, and behavior—that destabilize human exceptionalism and forward the posthumanist appreciation of plant life, as summarized, for instance, in plant neurobiologist Stefano Mancuso's book *The Revolutionary Genius of Plants*. Informed by integrative modes of science, critical posthumanism reconceptualizes life writing by calling into question "the autonomous self, the pact between author and reader, the foregrounding of the human" through an emphasis on relational-material Umwelten, or lifeworlds.[36]

Although concerned explicitly with human communities in the Caribbean, Édouard Glissant's thoughts on otherness help to elucidate the idea of phytography in the light of human-plant difference: "I thus am able to conceive of the opacity of the other for me, without reproach for my opacity for him. To feel in solidarity with him or to build with him or to like what he does, it is not necessary for me to grasp him. It is not necessary to try to become the other (to become other) nor to 'make' him in my image."[37] Decentering the sense of sight and the aesthetics of the visual, phytographical writing embraces the opacity of plants through attention to the complexities of sound, touch, taste, and smell in their lives.[38] With attention to the subjectivities of photosynthetic more-than-humans, phytography confers to plants the right to narrate their own stories. Given agential standing, plants thus become "coaxers of life narrative."[39] Extending Huff's posthumanist framing of life writing, phytography eschews stories of "progress in favor of those disrupting any linear narrative, telling, instead, a becoming together."[40] Privileging relational subjectivities, furthermore, phytography narrativizes a permeable "space where many beings—some of them readily visible, some not [for example, plants]—continually affect each other in contradistinction to the story of an individual human moving through and dominating space."[41]

Writing with, and in response to, plants in this way requires becoming conversant with their communicative modalities. A useful conceptual platform is phytosemiotics, or the study of sign processes within and between plants.[42] Phytosemiotics has been inspired by the research of biologist Jakob von Uexküll, who, in 1922, observed that the "comfortable calm" of plants conceals a bustle of communicative transactions from the human eye.[43] To be sure, the field of phytosemiotics regards language and signification in the vegetal world as functions of plant physiologies. From a phytosemiotic perspective, sign processes, including immunological responses and intercellular communication, express the plant's inner experiences of the world at the microscalar level of cells and tissues. In a related sense, Patrícia Vieira theorizes the notion of *phytographia*—or, literally, "plant writing"—as denoting "the specific modes in which the vegetal world is embedded in human cultural productions."[44] For Vieira, the concept signifies the permeability between literary representations of flora and "the imprints left in texts by the plants themselves."[45] As such, a plant-focused narrative is necessarily heteroglossic as vegetal and human forms of inscription interact and entangle.

In previous conceptualizations of vegetal language, furthermore, I have assayed the potential for poetic narrative to become "plant script" composed in response to "the endemic semiosis of vegetal life."[46] The term *plant script* signifies the nonverbal forms of expression specific to vegetal life and comprising, for instance, pheromonal transmissions, electrical signals, acoustic signatures, and corporeal articulations (the curling of tendrils in response to sunlight or the bifurcation of the tree branch following insect infestation).[47] For me, as for Vieira, plant script intermeshes with texts and processes of textual composition. Drawing from Bose's experimental work, this principle encompasses the autopoietic expressions of plants enacted by—rather than imposed on—them as agents in the world. In these two theoretical models of vegetal language, the enunciations of plant life contribute to the poiesis—the becoming, evolution, maturation—of botanical narratives themselves. These enunciations signify possibilities of growing, suffering, competing, aspiring, and other affective vegetal states that can emerge in narratives of plant intelligence wherein the human imagination is allowed to transcend the limits of rationalistic plant discourse and the stigma of anthropomorphism.

Plant Lives in Botanical Nonfiction: From Thoreau to Haskell

Unsettling humanistic compulsions within traditional life writing and engaging critically with ecological understandings of flora, phytography calls attention to the complexities of plants and plant-Umwelt interactions. Toward collaborative exchange with botanical life, phytographical writing also contests the stigma traditionally associated with anthropomorphism, or "the attribution of human properties to nonhuman entities."[48] Narrating individual plant lives in the context of their botanical confederations, this mode of more-than-human life writing often employs anthropomorphism to prompt readers to care more about vegetal others. On this note, Alexa Weik von Mossner observes that "anthropomorphism is inevitable when nonhumans become narrative agents in human stories, and yet there is a wide spectrum of how it has been employed."[49] As a response to such prevalence, she advocates *strategic anthropomorphism* as a means to engender empathy for, and relationality with, more-than-human subjects in narratives.[50] In a similar manner, Jane Bennett urges the "need to cultivate a bit of anthropomorphism—the idea that human agency has some echoes in nonhuman nature—to counter the narcissism of humans in charge of the world."[51] For Bennett, "a touch of anthropomorphism, then, can catalyze a sensibility that finds a world filled not with ontologically distinct categories of beings (subjects and objects) but with variously composed materialities that form confederations."[52] In reference to animal intelligence, moreover, ethologists Jesús Rivas and Gordon Burghardt proffer *critical anthropomorphism* as a foil to potentially harmful forms of anthropomorphism that become the implicit basis for drawing conclusions about the inner worlds of animals.[53]

A seminal example of phytographical nonfiction imbricating vegetal and human lives—and deaths—through the story of an individual plant persona is "Good Oak," the second, or February, section of the seasonally choreographed first part of Aldo Leopold's *A Sand County Almanac*, published originally in 1949.[54] The essay opens with the present-tense immediacy of an oak log burning in a fireplace during the Wisconsin winter: "The particular oak now aglow on my andirons grew on the bank of the old emigrant road where it climbs the sandhill."[55] In death, the tree retains narrative *voice*, not as a ventriloquization of human expression but as the registering of a specific material-affective "cluster of feelings triggered by life finding a way to announce itself."[56] Even as it is consumed by the fire, the oak remains an emplaced presence with an identity constituted in space and time. Rather than exclusively foregrounding biogeographic patterns, population distributions, and broadscale species interactions, phytography calls attention to the existences of particular plants vis-à-vis their places of inhabitation: the sandhill, the old road, the andirons. In Leopold's account, moreover, the oak corpus inscribes time's advance and, accordingly, temporalizes history and historical consciousness. The stump "shows 80 growth rings, hence the seedling from which it originated must have laid its first ring of wood in 1865, at the end of the Civil War."[57] Mediated by the tree body, the synchronization of cultural and botanical pasts is a form of strategic anthropomorphism. Rousing empathy for, and identification with, arboreal lives, Leopold figures the oak seedling as an offspring of a transformative yet immensely tragic period in US history.

A technique used by Leopold is to treat the death of the emplaced tree-persona as an ecologically and emotionally reverberative event that impacts myriad lives, including his own. For literary-studies scholar Erin James, this second section of *A Sand County Almanac* is a "biography of the individual oak" predicated on "collaboration between plant and human language" toward the celebration of the life and death of a tree.[58] Moving away from reductive generalizations and toward affective regard for a singular plant entity, Leopold tells us that a lightning bolt "put an end to wood-making by this *particular* oak."[59] For Leopold, delineating the oak's experiences—in dialogue with the embodied oak itself—necessitates attentiveness to the tree's decline and demise. Narrative memory in "Good Oak" coheres with the corporeality—and mortality—of the tree as "the trunk showed a long spiral scar of barkless sapwood, a foot wide and not yet yellowed by the sun."[60] Although triggering mourning in the narrator, the loss of the old oak invariably leads to ecological renewal as "a dozen of its progeny standing straight and stalwart on the sands had already taken over its job of wood-making."[61] In a marked shift of course, the narrative then synchronizes the act of sawing with the passage of interlinked human-vegetal temporalities. The blade cuts "stroke by stroke, decade by decade, into the chronology of a lifetime, written in concentric annual rings of good oak."[62]

Thus, the composition of Leopold's phytographical narrative comes to approximate the speaker's very movements through the oak corpus. The upshot is a call-and-response text written-with the arboreal body and organized around the present-tense refrain "rest! cries the chief sawyer, and we pause for breath."[63] The metempsychotic capacity of the oak enlivens—literally, inspirits—multispecies regeneration following its incineration on the andirons. The ashes "will come back to me again, perhaps as red apples, or perhaps as a spirit of enterprise in some fat October squirrel, who, for reasons unknown to himself, is bent on planting acorns."[64] James suggests that Leopold's phytography can be understood "as a delicate dance between human imagination and plant signification," in which the "oak's material language … dictates the pace and focus of his narration."[65] Maintaining careful, strategic anthropomorphism—without collapsing into the distanced, objective mode privileged in traditional scientific discourse—Leopold's mediation of human-vegetal intersectionality takes into consideration the forms of experience and modes of signification of the tree. As postulated by Rivas and Burghardt, critical anthropomorphism of this kind can help to overcome—rather than reinforce or perpetuate—biases against more-than-human beings by stimulating awareness of the inherent relationality between all lives nested in a place. As such, the oak is not a vegetal subordinate—rendered and reduced in prose—but an active contributor to, and autonomous agent in, the phytographical progression. This is so even in the oak's death; the essence of the tree remains in the text just as it does in the fat squirrel, planted acorns, and red apples—of then and of now.

First appearing in 1862, American naturalist Henry David Thoreau's essay "Wild Apples" presents another phytographical example, distinct for its highly nuanced attention to multisensorial interaction with apple-tree personae.[66] Throughout his narrative corpus and particularly in works such as *Faith in a Seed*, Thoreau developed a transdisciplinary, culturally inflected model of botany contingent on local, individual plant personae while eschewing anthropocentric ontologies of plants.[67] In Thoreau's botanical prose, the trees write-back into the text—with the human author and reader as mediators—via their diverse haptic, olfactory, and gustatory significations. The opening passages signal the interplay of seasons and sensorialities that is characteristic of Thoreau's approach: "early apples begin to be ripe about the first of August, but I think that none of them are so good to eat as some to smell."[68] Comparable to the first part of Leopold's *A Sand County Almanac*, Thoreau's narration of botanical lives creates intimacy with individual vegetal beings as they change—seed, fruit, flower, sprout, mature, decline—across time spans and seasons. More specifically, "Wild Apples" emphasizes smell- and taste-related memory as a means to engender rapport with plants while imbuing human language with vegetal forms of enunciation: "The fragrance of some fruits is not to be forgotten, along with that of flowers."[69] Vivid prose particularizes one tree in space and time as Thoreau remembers it: "Going up the side of a cliff about the first of November, I saw a vigorous young apple tree which, planted by birds or cows, had shot up amid the rocks and open woods there and had now much fruit on it."[70] The young tree was "a rank, wild growth, with many green leaves on it still, and made an impression of thorniness."[71] Its fruit—some dangling on branches, some scattered on the ground, some bowled down the hill by gravity—appeared green and inedible.

Phytographical writing, such as Thoreau's "Wild Apples" and Leopold's "Good Oak," generates a porous body-text-plant-Umwelt where, in keeping with Huff's assertion, "many beings—some of them readily visible, some not—continually affect each other."[72] To be certain, Thoreau championed the lives of vegetal beings—including the young, rank, and wild apple—that might otherwise remain opaque, invisible, and unappreciated, except to more-than-human members of its community: "The day was not observed when it first blossomed nor when it first bore fruit, unless by the chickadee."[73] Later, the essay narrates a recollection of encountering a crab apple at Saint Anthony Falls in Minnesota at the northern limit of its range: "I succeeded in finding it about eight miles west of the Falls; touched it and smelled it, and secured a lingering corymb of flowers for my herbarium."[74] The nonfiction of Thoreau and Leopold is predicated on sustained interest in vegetal personae through time and over the seasons. In slight contrast, contemporary botanical prose tends to integrate individual plant lives as rhetorical devices for delineating broadly sweeping natural-history and ecological narratives.[75]

As a case in point, the tendency to overlook particular plants (as *lives*) in favor of generalized plants (as *life*) surfaces in the essay "A Question of Character" from *The Hidden Life of Trees* by Wohlleben.[76] On first glance, most readers would think that the idea of *plant character* aligns perfectly with the examination of individual vegetal lives, such as the subjective inner worlds of Leopold's oak and Thoreau's apple. Wohlleben's chapter indeed begins on a promising note, with Thoreauvian-Leopoldian specificity: "On the country road between my home village of Hümmel and the next small town in the Ahr valley stand three oaks. They are a commanding presence out in the open fields, and the area is named in their honor."[77] The oak trio is an emplaced presence that reverberates within its biocultural milieu. Each oak, moreover, exhibits unique behavior because of its "innate characteristics."[78] Wohlleben elaborates that "whereas the oak on the right is already turning color, the middle one and the one on the left are still completely green."[79] However, the author's empathy-engendering particularization of individual oaks within the grove soon veers in a different direction. Leaving the oak trio behind by the end of the second paragraph of the chapter, Wohlleben suggests tantalizingly that "the timing of leaf drop, it seems, really is a question of character."[80] The narrative then plunges headlong into a synopsis of photosynthetic processes, the anatomy of deciduous trees, and mechanisms used by oaks to ward off fungal invasions. This descent from the particularized to the essentialized—from *plant lives* to *plant life*—is marked by the following kinds of declaratives: "Trees cannot anticipate the coming winter. They don't know whether it is going to be harsh or mild. All they register are shortening days and falling temperatures."[81]

A comparable progression—from *lives* to *life* —also pervades Haskell's rendering of the balsam fir in the second chapter of *The Songs of Trees*. An epigraph signifies the geographical precision of the author's encounter with the tree: "Kakabeka, north-western Ontario, 48°23'45.7 N, 89°37'17.2 W."[82] The evocative initial passages of the chapter immerse the reader in a fir's seasonality, sensoriality, and physicality: "In summer the slate blue cones were clenched shut. Copious dribbles of resin kept away birds and squirrels. Now, in October, the cones have browned and the dried resin has fallen. Scales have eased apart to reveal stacks of thin, translucent paper. A flick of wind shatters the cone with a gentle snap and hiss, then paper kites stream away, some carried high, others spinning to the ground. Each kite has a traveler clinging to its base, a balsam fir seed barely thicker than the paper that carries it."[83] Like Wohlleben's treatment of the oak, however, Haskell's highly particularized account of the fir gravitates toward a natural-history mode that privileges the collective while at the same time tending to obscure the individual. In contrast to Wohlleben's narrative of the oak trio, Haskell's narrativization oscillates, at times, between the balsam-fir personae and the generalities of its genus-species. Notwithstanding the individual character of the tree—within its lifeworld but also as an Umwelt for other intelligent creatures, including chickadees—the fir becomes an essentialized category of nonhuman beingness through declaratives including, "Some plant responses are long term, such as the growth of branches into light or roots into fertile soil. Plant architecture is not a haphazard affair but is the result of constant assessment and adjustment as conditions change."[84] What strikes me as diminished, to varying degrees, in the narratives of Haskell, Wohlleben, and other contemporary botanical non-fiction writers is a balanced, even dialogical sense of the plant in focus as "at once the most singular and the most general being."[85]

Phytographical Poetry: Letting Plants Tell Their Own Stories

Encompassing nonfiction and poetry, phytography as conceptualized in this article destabilizes the boundary between biography and autobiography as the voices—articulations, enunciations, gesticulations—of plants figure into the form and content of narratives. I postulate that, to some extent, all phytographical writing absorbs the inflections of botanical beings, and therefore an element of coauthorship is inescapable. From a posthumanist standpoint, plant life writing is inherently a polylingual, heteroglossic "becoming together" as vegetal and human linguae interleave.[86] The principles of writing-with plants collaboratively and plants writing-back into texts figure into examples of contemporary *phytographical poetry*, which I define as the writing of botanical lives in verse. A cross section of poems—on vegetal sensorialities, first-person flower voices, plant-insect co-vocalizations, and experimental poem-transcripts derived from tree interviews[87]—is illustrative of phytographical poetry as human authors *writing-with* plants in response to the material interpolations of vegetal life.

Australian poet Andrew Lansdown's "A Few Weeks Later I Return to Find," from his 1979 collection *Homecoming*, focuses on multisensorial engagement with a balga tree (*Xanthorrhoea preissii*), endemic to the southwest corner of Western Australia. Instead of attempting to create a narrative portrait of the individual plant's life—from germination to maturation—Lansdown's poetic phytography elaborates an exact moment of nonverbal poet-balga interchange. A strong undertow in the poem, furthermore, is the dual role of botanical science in both disclosing and obscuring botanical lives. Thus approximating within the text the friction between scientific and sensory epistemologies of plants, the account opens with a visual anatomization of the *Xanthorrhoea preissii* flower, tracking between taxonomic and poetic language:

> Centred in the *stamens*,
> the shorter *stylus*—surrounded,
> and at times, 'covered
> by a glistening glob of transparent nectar
> which, in turn, was caught in the cup-pit
> of the six guardian *stamens*.[88]

Following this morphological snapshot, the poem begins to elicit a sense of the *Xanthorrhoea preissii* as a particular vegetal being representative of its species. Grounded in multisensorial dialogue, the poet-plant exchange reaches a crescendo when

> I thought each flower had mysteriously
> caught last night's dew,
> so I put my tongue to it
> (Descartes would not have approved) to see:
> it was a powerful, honey-thick
> nectar. The odour was a heavy
> sweetness. I wiped the pollen from my nose.[89]

Lansdown's description of the tonguing of the nectar evokes a moment of joyous, even erotic, intersubjectivity between poet and plant. The balga expresses its fertile being-in-the-world via the material registers of the nectar: powerful, heavy, sweet, and honey-thick. Hence, gustatory, olfactory, and tactile transactions between the poet and the balga disrupt the detachment of the Cartesian *cogito ergo sum*. The narrator participates in—rather than perceives from a distance—the material signifiers of vegetal transformation over time. In this regard, the balga is not a static, passive object but a mutable, intelligent phenomenon capable of writing its own life through its own bodily articulations. As such, Lansdown's narrative can be read as a collaborative inscription at the human-plant conjunction.

In Louise Glück's Pulitzer Prize-winning volume *The Wild Iris*, published in 1992, plants also write their own lives in dialogue with the narrator. In the first-(plant)person voice, garden flowers express their inner worlds—emotions, memories, and states of consciousness—through vegetal script facilitated by the poet. The titular poem, "The Wild Iris," inscribes the sapience of the flower. In their perennialism, irises archive the traces of seasonal deaths and resurrections. The conscious flower bulb remembers its interment as it anticipates spring's arrival:

> Overhead, noises, branches of the pine shifting.
> Then nothing. The weak sun
> flickered over the dry surface.
>
> It is terrible to survive
> as consciousness
> buried in the dark earth.[90]

To be sure, Glück's stanzas cohere with current understandings of visual, sonic, and proprioceptive perception in plants.[91] Contrary to Friedrich Nietzsche's assertion that memory—but not consciousness—prevails in the vegetal kingdom, Glück figures the iris bulb as a plenum of intellective activity. The expansiveness of vegetal consciousness casts in sharp relief the limited capacity of human recollection, emphasized by the iris-speaker's characterization of—that is, its writing-back to—the poet-gardener as "you who do not remember | passage from the other world."[92] The flower's blunt declarations reflect its individual character and personality within its life arc. The enactment of vegetal cognition in "The Wild Iris" entails the narration of plant voice as constituted by the embodied significations with which the shamanic poet-gardener writes in response. Glück's botanical poetics remind us that writing-with flora is predicated on receptivity to their forms of writing-back (their voices, gestures, expressions, interpolations). Narrating how our experiences imbricate with those of plants, phytography indeed emerges at the intersection of vegetal and human languages. In short, writing botanical lives in this way demands protracted attention to the intelligence and creativity of plants, as well as their beauty and uses.

Whereas the phytographical poetry of Glück and Lansdown employs the lyric mode, Jody Gladding's *Translations from Bark Beetle* from 2014 pushes boldly into the province of human-plant-insect auto|biography and multispecies coauthorship. Part ephemeral and found art, part experimental poetics, *Translations* comprises two kinds of phytographical poetry. Both situate the vibrant materiality of plants as the substratum for polylingualism in human and more-than-human worlds. In the first, Gladding transliterates "notations" etched in trees by bark beetles and, in the second, she engraves her own poems on natural-media slate.[93] The beetle engravings—the material significations of entangled lives—supply "both a shape and a language for each poem."[94] The book features the beetles' "encoded glyphs," minimally mediated in the form of graphite rubbings.[95] Narrating the covocalization of insect, tree, and writer, the polyphonic collection "calls us to a deeper communion—*true collaboration*—between art and the more-than-human world."[96]

Translations epitomizes Miller's argument that, in autobiography and other kinds of life writing, "the relational is not optional" within a "web of entanglement."[97] This assertion concretizes in "Engraver Beetle Cycle" and, in particular, the poem's idiosyncratic use of black dots to call out the linguistic reduction of subjectivities to pronouns:

> Yo• can only travel in one direction
> but turn again with m• there love
> sap in the chamber
> red the friable
>
> ("Spending Most of their Time in Galleries,
> Adults Come into the Open on Warm Sunny
> Days")
>
> through work the quietly
> puncture begins in a dark
> if not there's no
> telling
> (rue mores of light and lying)
> some have remained here burrowed.[98]

The poem integrates human, arboreal, and coleopteran enunciations. Not only a medium for the articulations of other beings, the tree is an active participant and contributing voice. In an interview, Gladding explains that the compositional process becomes collaborative as the poems begin to synthesize "what I understand about the way bark beetles make poems with what I understand about making poems myself."[99]

Like Gladding's *Translations*, Burk's *Tree Talks*, appearing in 2016, presents a series of encoded glyph-like patterns evocative of the inscription of plant neuroactivity on the smoked-glass plate of Bose's crescograph. *Tree Talks* comprises poem-transcripts of "unstructured interviews with 8 Southern Arizona trees," including pines and willows, who acted as participants in the poet's field-based study of "ethics, environment, politics, communication, and failure to communicate [between species]."[100] The quasi-social-science approach is at once an earnest enactment of "multispecies ethnography," a "mock-ethnographic study," and "a satire on research methodology."[101] The poem-transcript titles consist of Latin and common botanical names followed by the location and time of the interview, as in "*Salix gooddingii* (Goodding Willow) / Patagonia-Sonoita Creek Preserve / 2010.07.24 11:50)."[102] Onomatopoeia, capitalizations, brackets, parentheses, and ellipses, along with exclamation marks, full stops, semi-colons, em-dashes, and other punctuation, schematize the interviews yet signify "the inability of semiotics to decode environmental systems or point to the postconceptual."[103] The questions posed to the arboreal interviewees attempt to elicit insights into their experiences of time, place, ecology, each other, and the poet herself. In an interview with a One-Seed Juniper, the poet asks, "I was last here a little more than a year go. What's been happening around here since then?" To which the tree begins its reply with,

> [thunder] S:
> BrrrrrrNNNNNNNNNNNNNNNNNNNNNNNN.[104]

Situating the individual subject within its heteroglossic collective, *Tree Talks* provokes the reconsideration of human tenets of voice and communication. The work affirms that writing the lives of plants in dialogue with them is predicated on a willingness to negotiate vegetal ontologies while remaining open to the potential for moments of interspecies breakthrough.

Writing-With Plants: "Gorge: Scriptorium | Tree | Excubitorium"

I turn now to the phytographical poetry I have been composing in collaboration with the plants of the Northern Tablelands of New South Wales, Australia, a region known for its unusual elevation, plunge waterfalls, four seasons, and botanical diversity.[105] Largely cleared of its original vegetation since European colonization, the region comprises a network of gorges around which one of the earliest Anglo-European conservation systems in Australia developed. Although the plateau has been heavily modified by pastoral activity, the gorges remain sanctuaries for resilient plants. "Gorge: Scriptorium | Tree | Excubitorium" is a sequence exploring the idea of writing botanical lives in conversation with the plants themselves. Although some of the detail of the poems will remain opaque—especially for readers outside the Tablelands—"Gorge" represents an experiment in coauthoring with plants through a variety of techniques, including burying poems, mimicking plant sounds, and writing from the first-(plant)person perspective. The sequence reflects my immersion in the New England plantscape and development of close relations over time with particular plants and species.

In the title of the first part of the sequence, my use of the Latinate noun *scriptorium*—a place where monks transcribe texts—connotes the austerity, isolation, and ecological urgencies of the region's gorges. Yet, the term also suggests that the chasmic landscapes are loci of more-than-human linguae, in which plants write the scripts of their lives in the material strata of themselves and others—letters addressed to us, each other, and their divinities. I understand the gorges, then, as fundamentally linguistic topoi, constituted by vocabularies neither fully recognized nor comprehended by humankind. In the radical-landscape mode theorized by Harriet Tarlo,[106] the poems of the first part, "Scriptorium," attempt to push boundaries of poetic form while remaining topographically specific. Precise locations and Global Positioning System coordinates supplied in Courier-font headlines serve ambiguously as titles locating the poems in a geographic sense while serving as rubrics for cataloguing site visits in non-poetic language. As in Gladding's *Translations* and Burk's *Tree Talks*, the translation between human and more-than-human languages proves replete with difficulties and uncertainties.

The second part bears the simple name "Tree." This de(re)generative midsection of "Gorge" consists of variations on the Petrarchan sonnet *written-with* chasms, plants, insects, soil, elements—cold, warmth, moisture, dryness—and the processes of decay. I chose the Petrarchan sonnet as the parodic vehicle for plant script in order to destabilize the humanistic tradition of the form. Centered on the page as textual bodies, the sonnets adhere to, yet often deviate from, the conventional Petrarchan form. "Tree" incorporates concrete visual arrangements, archaic language, taxonomic allusions, and *glossolalic impulses* to provoke linguistic dissolution and convergence. Toward the possibility of writing-with plants in response to their writing-back, I plant, compost, digest, and seed poems at New England gorges. Human-vegetal reciprocity emerges through the ceremonial earthing and unearthing—*gorging and disgorging*—of sonnets and poem fragments. An assemblage of seen and unseen chasm-dwellers works over the source sonnets. They contribute their own terms, connotations, inflections, syntax, marginalia, elisions, and deletions, which I then integrate into subsequent sonnets. The open-ended cycle fluctuates with seasonal conditions, soil moisture, insect activity, microbial digesting, human memory, and other processes.

An untitled sonnet beginning "Nor am eye mere spectacle" emerged when source poems fused to produce composite lines. Additional weeks of de(re)composing—and coauthorial fiddling—with a gorge hakea resulted in a poem opening with "A depth of death I am sans abandon," included below. Shunning the third-person narrative voice, the sonnets make use of the first-(plant)person mode of address. Endowing Tableland plants with cognition, perception, and emotion, the heteroglossic verses turn the critical gaze of canny plants back toward the human writer and reader. As such, the sonnet cycle joins contemporary phytographical work, such as Glück's *The Wild Iris*, narrated from the standpoint of vegetal beings and based on a view of botanical subjects as inherently conscious. As the plants express their interior lives liberated from the distancing function of the third person, the sonnets come to represent ever-evolving collaborative forays into the percipience of botanical lives:

A depth of death I am sans abandon,
as slanted sun soothes verbforms in tension,
strokes blossom orchid hyphae elisions.
Were dying greater than remiss of one?
Were living lightened by the cease of none?
The frosted morning foisted a sudden
falling, nay to earth, tho' towards a coven
of boulders. I hardened to their contours.
To learn, I had to spruik their speech in death,
advised as such to snatch tongue of lichen.
Then you would grip a voice deprived of breath
granted the grammar from here to liken.
You discern my murmur within this cleft.
We transit to death through lives alike in.[107]

Phytographical composting is an experimentation with more-than-human editing in order to introduce the voice(s) of vegetal nature into the poems. The degradation of the sonnets underscores their material basis and the heterogeneous relations of poems within Umwelten. Composting is also symbolic of poems as seeds—of hope, renewal, inspiration, and transformation. In this context, author-activist Rebecca Solnit comments that "writers understand that action is seldom direct. You write your books. You scatter your seeds. Rats might eat them, or they might just rot … some seeds lie dormant for decades because they only germinate after fire."[108] The sonnet-seeds of "Tree" symbolize the restorative potential of human-plant alliances in a bioregion that has been widely devastated by colonization in the past two hundred years.

The title of the third section, "Excubitorium," refers to a place of poetic vigil and represents the coda of the "Gorge" sequence. In a church gallery, an excubitorium was the area "where public watch was formerly kept at night on the eve of a festival" and—in reference to medieval monasteries—the term denotes "an apartment for night watchers whose duty it was to call monks to their nocturnal devotions."[109] In relation to phytographical poetry, excubitorium evokes the prayerful vigil that plants hold over plateau ecosystems, as the photosynthetic fundament on which life depends. The obscure Latinate term conjures the monastic austerity of the chasms while also inflecting the technical language used to assign plants to families, genera, species, and other categories. This part of "Gorge" comprises three sonic movements organized around the three main tree species populating the rim of Dangar's Gorge near the town of Armidale: Gorge Wattle (*Acacia ingramii*), Gorge Bertya (*Bertya ingramii*), and Bulloak (*Allocasuarina luehmanii*). My phytographical method involved recording the soundscapes of the trees. I then composed the poems in response to their sonic environments:

Windriffed bloodtide whisk—*crunk grasp flasp flisk*

heartbeat crests gorge susurrus then smashes over us
in arterial tides murmuring, slurs and swells saccadic,
cardiac fibrillation then zipper twill. *What meditative
gunk is this? What do my tufts of tillandsia whisper?*

Gorge rim soon to flower—*scrunch*—flick of Bic lighter,
plastic click, feedback underheel, thump on drum skin,
crescendo of interstitial rasp, *somebody's about to gasp*,
polite formalities then interview ends, hasty handclasp.
My voice is a heteroglossic bird before you [digital|*crash*].

At first, I was calm, but your inner tumult overtook me,
I became withdrawn, reticent—*cough groan ugh*—crest-
fallen, blind(in)sided, introspectively-drawn, revenant far
blown-off, parasail glided, migraine lumbered, alone on
this lip with the godless, quivering, grousing, muttering.

Lento, adagio, *I fall*, wind-knocked, -crisped, -asthmatic,
in octaves of chalkboard scratch, freight trains rumbling
on tracks, thunderclaps of waves over boulders, fermata,
decrescendo, *I speak as a collective zephyr breath [yes I!]*
fortissimo, ethereal vi(r)gi(l)n in G minor—diminuendo.[110]

There are instances where the words correspond to the audio signature of each species and become mimetic of the soundscapes of gorge plants. Although a playful enactment of the idea of plants writing-back, the poems do reflect theoretical developments in bioacoustics, an area of scientific research indicating that plants have voices, though radically unlike our own. Phytoacoustic research, for example, indicates that different species and individual plants emit specific sound signatures that enable them to make decisions and communicate with other organisms.[111] The ecological function of sound implies the presence of intelligence in the plant, which has been regarded in the Western intellectual tradition as the polar opposite of the animal—as passive, mute, and deprived of cognition.

Conclusion: Phytography as Posthumanist Plant Life Writing

In critical dialogue with botanical science, phytography on the whole offers insights into the lives of plants. As posthumanist life writing predicated on collaboration with, and receptivity to, plants, phytography also prompts the reassessment of the human preeminence within traditional auto|biography. It does so without privileging botanical communities and ecological systems over individual plants and vegetal personae. The nonfiction of Thoreau, Leopold, and, to various extents, Haskell and Wohlleben and the poetry of Lansdown, Glück, Gladding, and Burk demonstrate phytographical writing's nuanced attention to plant intelligence, behavior, corporeality, temporality, seasonality, emplacement, language, historicity, and mortality. In the phytographical model I have theorized, "the author and reader, and the relationship between these two, would not hold a privileged position, nor would the discrete entity of a text. Subjectivity would not be elided to emphasize human exceptionalism, nor would human language assume centrality."[112]

Phytography, as delineated here, hinges on the dialogical interplay between *writing-with* plants (collaboratively) and remaining vigilant to their *writing-back* (those material inscriptions through which they express their lives regardless of our textual or linguistic intercessions). I have suggested that the being-in-the-world of plants—increasingly articulated by studies of vegetal behavior, cognition, and communication—is a form of life writing in itself. The challenge for the phytographical writer is to recognize these expressions of language wherever plants are encountered. Nonetheless, in closing, I stress that the overwhelming emphasis within the widely ranging phytographies presented in this article is the arboreal. Although the work of Thoreau and Glück presents exceptions—the former, for instance, wrote also about the lives of cranberries and grapes—small trees, shrubs, and herbaceous plants are alarmingly absent. This could reflect the difficulty of engaging multisensorially with the lives of non-trees alongside the broader arborealization of Western thought called out by Gilles Deleuze and Félix Guattari in their theorization of the rhizome.[113] As all kinds of plants—not only trees—write the narratives of their own lives, the onus is on the human author to sense, listen, wait, and allow their voices to come forth.

University of New England and Southern Cross University

Notes

1. See, for example, Chamovitz, *What a Plant Knows*; Haskell, *The Songs of Trees*; Mabey, *The Cabaret of Plants*; and Wohlleben, *Hidden Life of Trees*. Published in 2014 and set in southeast Queensland, Greer's *White Beech* should also be noted as an important contemporary example of plant-attentive place-based memoir weaving between history, botany, and feminist politics.
2. Wohlleben, *Hidden Life of Trees*, 1–5, 6–13, 19–24, 79–84.
3. Haskell, *The Songs of Trees*, 37.
4. See, for example, Karban and Orrock, "Judgment."
5. Audus, "Roots of Absurdity"; Baluška and Mancuso, "Plants and Animals," 285; and Green, *Secret Life of Plants*.
6. Bose, "Literature and Science."
7. Haskell, *The Songs of Trees*, 31–58.
8. Haskell, *The Songs of Trees*, 39.
9. Marder, *Plant-Thinking*, 183.
10. Huff, "After Auto, after Bio"; Huff and Haefner, "His Master's Voice."

11. Bennett, *Vibrant Matter*, xvi; Weik von Mossner, *Affective Ecologies*, 132.
12. Marder, *Plant-Thinking*, 183.
13. Baluška et al., "Neurobiological View of Plants," 20.
14. Gagliano, Ryan, and Vieira, "Introduction."
15. Dretske, "Machines, Plants and Animals," 25; Marder, *The Philosopher's Plant*, 148.
16. Wickens, *Economic Botany*, xiii. Even contemporary botanical writers have become ensnared in vegetal utilitarianism—valuing flora narrowly vis-à-vis its usefulness. As a case in point, Mabey's *Plants with a Purpose*, published in 1977, focuses on "the ways in which the commoner wild plants of Europe and North America have, and can, be put to fruitful and enjoyable household use" (12).
17. Mancuso and Viola, *Brilliant Green*.
18. Smith and Watson, *Reading Autobiography*, 4.
19. Ryan, *Plants in Contemporary Poetry*, 7–8.
20. See, for example, Haskell, *The Songs of Trees*, 31–58; Leopold, *A Sand County Almanac*, 8–17; and Thoreau, *Wild Fruits*, 74–92.
21. Smith and Watson, *Reading Autobiography*, 201–203.
22. Ryan, *Plants in Contemporary Poetry*, 7–10.
23. Huff, "After Auto, after Bio," 281.
24. Husserl, *Pure Phenomenology*, 52–55.
25. Miller, "The Entangled Self," 544.
26. Huff and Haefner, "His Master's Voice," 153.
27. Huff and Haefner, "His Master's Voice," 154.
28. Huff, "After Auto, after Bio," 279.
29. Huff and Haefner, "His Master's Voice," 159.
30. Arana, "Biography and Poetry," 116.
31. Arana, "Biography and Poetry," 117.
32. Abbs, "Autobiography and Poetry," 82.
33. Abbs, "Autobiography and Poetry," 81.
34. Abbs, "Autobiography and Poetry," 82.
35. Huff and Haefner, "His Master's Voice," 165.
36. Huff, "After Auto, after Bio," 279.
37. Glissant, *Poetics of Relation*, 193.
38. Huff, "After Auto, after Bio," 279.
39. Huff, "After Auto, after Bio," 280.
40. Huff, "After Auto, after Bio," 280.
41. Huff, "After Auto, after Bio," 280.
42. Krampen, "Phytosemiotics."
43. Qtd. in Krampen, "Phytosemiotics," 266.
44. Vieira, "*Phytographia*," 218.
45. Vieira, "*Phytographia*," 218.
46. Ryan, "Poetry as Plant Script," 127.
47. Gagliano, Ryan, and Vieira, "Introduction."
48. Rivas and Burghardt, "Crotalomorphism," 9.
49. Weik von Mossner, *Affective Ecologies*, 107.
50. Weik von Mossner, *Affective Ecologies*, 132.
51. Bennett, *Vibrant Matter*, xvi.
52. Bennett, *Vibrant Matter*, 99.
53. Rivas and Burghardt, "Crotalomorphism," 10.

54. Leopold, *A Sand County Almanac*, 8–17.
55. Leopold, *A Sand County Almanac*, 8.
56. Watts, "Coda," 259.
57. Leopold, *A Sand County Almanac*, 8.
58. James, "What the Plant Says," 267.
59. Leopold, *A Sand County Almanac*, 10; emphasis added.
60. Leopold, *A Sand County Almanac*, 10.
61. Leopold, *A Sand County Almanac*, 10.
62. Leopold, *A Sand County Almanac*, 10.
63. Leopold, *A Sand County Almanac*, 15 and elsewhere.
64. Leopold, *A Sand County Almanac*, 17.
65. James, "What the Plant Says," 268–269.
66. Thoreau, *Wild Fruits*, 74–92.
67. Ryan, *Green Sense*, 13–40.
68. Thoreau, *Wild Fruits*, 74.
69. Thoreau, *Wild Fruits*, 74.
70. Thoreau, *Wild Fruits*, 78.
71. Thoreau, *Wild Fruits*, 78.
72. Huff, "After Auto, after Bio," 280.
73. Thoreau, *Wild Fruits*, 78.
74. Thoreau, *Wild Fruits*, 80.
75. Chamovitz, *What a Plant Knows*; Haskell, *The Songs of Trees*; Mabey, *The Cabaret of Plants*; Mabey, *Plants with a Purpose*; Wohlleben, *Hidden Life of Trees*.
76. Wohlleben, *Hidden Life of Trees*, 151–154.
77. Wohlleben, *Hidden Life of Trees*, 151.
78. Wohlleben, *Hidden Life of Trees*, 151.
79. Wohlleben, *Hidden Life of Trees*, 152.
80. Wohlleben, *Hidden Life of Trees*, 152.
81. Wohlleben, *Hidden Life of Trees*, 152.
82. Haskell, *The Songs of Trees*, 32.
83. Haskell, *The Songs of Trees*, 32.
84. Haskell, *The Songs of Trees*, 36.
85. Marder, *Plant-Thinking*, 183.
86. Huff, "After Auto, after Bio," 280.
87. Lansdown, *Homecoming*; Glück, *The Wild Iris*; Gladding, *Translations from Bark Beetle*; Burk, *Tree Talks*.
88. Lansdown, *Homecoming*, 8, lines 11–16; emphasis added.
89. Lansdown, *Homecoming*, 8, lines 17–23.
90. Glück, *The Wild Iris*, 1, lines 5–10.
91. Chamovitz, *What a Plant Knows*.
92. Glück, *The Wild Iris*, 1, lines 16–17.
93. Bervin and Gladding, "Three Dimensions."
94. Milkweed Editions, "*Translations from Bark Beetle*."
95. Bendall, "Secret Inscriptions."
96. Milkweed Editions, "*Translations from Bark Beetle*"; emphasis added.
97. Miller, "The Entangled Self," 544.
98. Gladding, *Translations from Bark Beetle*, 6.
99. Bervin and Gladding, "Three Dimensions."
100. Burk, *Tree Talks*, preface.

101. Kirksey and Helmreich, "Emergence of Multispecies Ethnography"; Russo, "Listening-Being"; Gessner, "The Synecological Poem."
102. Burk, *Tree Talks*, 21.
103. Gessner, "The Synecological Poem."
104. Burk, *Tree Talks*, 31.
105. Atkinson et al., *High Lean Country*.
106. Tarlo, "Introduction."
107. Ryan, "Gorge."
108. Solnit, "Acts of Hope."
109. Harris, *Dictionary*, 57.
110. Ryan, "Gorge."
111. Gagliano et al., "Tuned In."
112. Huff, "After Auto, after Bio," 281.
113. Deleuze and Guattari, *A Thousand Plateaus*.

Acknowledgments

The author is grateful for the support of the issue editors, Jessica White and Gillian Whitlock. Thanks also to Stuart Cooke for suggesting Burk's *Tree Talks*.

Disclosure Statement

No potential conflict of interest was reported by the author(s).

Works Cited

Abbs, Peter. "Autobiography and Poetry." In *Encyclopedia of Life Writing: Autobiographical and Biographical Forms*, edited by Margaretta Jolly, 81–82. London: Fitzroy Dearborn, 2001.

Arana, R. Victoria. "Biography and Poetry." In *Encyclopedia of Life Writing: Autobiographical and Biographical Forms*, edited by Margaretta Jolly, 116–117. London: Routledge, 2001.

Atkinson, Alan, et al., eds. *High Lean Country: Land, People and Memory in New England*. Crow's Nest, NSW: Allen and Unwin, 2006.

Audus, Leslie. "Roots of Absurdity." *New Scientist* 64, no. 919 (1974): 207.

Baluška, František, and Stefano Mancuso. "Plants and Animals: Convergent Evolution in Action?" In *Plant-Environment Interactions: From Sensory Plant Biology to Active Plant Behavior*, edited by František Baluška, 285–301. Berlin: Springer, 2009.

Baluška, František, et al. "Neurobiological View of Plants and Their Body Plan." In *Communication in Plants: Neuronal Aspects of Plant Life*, edited by František Baluška, Stefano Mancuso, and Dieter Volkmann, 19–35. Berlin: Springer, 2006.

Bendall, Molly. "Secret Inscriptions." *Free Verse* 25 (2014). http://freeversethejournal.org/issue-25-2014-molly-bendall-on-jody-gladding/

Bennett, Jane. *Vibrant Matter: A Political Ecology of Things*. Durham, NC: Duke UP, 2010.

Bervin, Jen, and Jody Gladding. "Three Dimensions: Jody Gladding on Translation, the Sources of Language, and How Beetles Can Speak of Longing." *Poetry Foundation*, August 20, 2014. https://www.poetryfoundation.org/articles/70136/three-dimensions

Bose, Jagadish Chandra. "Literature and Science." In *Sir Jagadis Chunder Bose: His Life and Speeches*, edited by Bryan Ness, 79–101. Charleston, SC: BiblioBazaar, 2007.

Burk, Wendy. *Tree Talks: Southern Arizona*. Fort Collins, CO: Delete, 2016.

Chamovitz, Daniel. *What a Plant Knows: A Field Guide to the Senses*. Brunswick, VIC: Scribe, 2016.

Deleuze, Gilles, and Félix Guattari. *A Thousand Plateaus: Capitalism and Schizophrenia*. Translated and edited by Brian Massumi. Minneapolis, MN: U of Minnesota P, 1984.

Dretske, Fred. "Machines, Plants and Animals: The Origins of Agency." *Erkenntnis* 51, no. 1 (1999): 19–31.

Gagliano, Monica, et al. "Tuned In: Plant Roots Use Sound to Locate Water." *Oecologia* 184, no. 1 (2017): 151–160.

Gagliano, Monica, John C. Ryan, and Patricia Vieira. "Introduction." In *The Language of Plants: Science, Philosophy, Literature*, edited by Monica Gagliano, John C. Ryan, and Patricia Vieira, vii–xxxiii. Minneapolis, MN: U of Minnesota P, 2017.

Gessner, Michael. "The Synecological Poem: *On Tree Talks: Southern Arizona* by Wendy Burk." *Kenyon Review*, 2016. https://www.kenyonreview.org/reviews/tree-talks-southern-arizona-by-wendy-burk-738439/

Gladding, Jody. *Translations from Bark Beetle: Poems*. Minneapolis, MN: Milkweed, 2014.

Glissant, Édouard. *Poetics of Relation*. Translated and edited by Betsy Wing. Ann Arbor, MI: U of Michigan P, 2010.

Glück, Louise. *The Wild Iris*. New York, NY: HarperCollins, 1992.

Green, Walon. *The Secret Life of Plants*. Los Angeles, CA: Infinite, 1978.

Greer, Germaine. *White Beech: The Rainforest Years*. London: Bloomsbury, 2014.

Harris, Cyril. *Illustrated Dictionary of Historic Architecture*. Mineola, NY: Dover, 1983.

Haskell, David George. *The Songs of Trees: Stories from Nature's Great Connectors*. Carlton, VIC: Black, 2017.

Huff, Cynthia. "After Auto, after Bio: Posthumanism and Life Writing." *a/b: Auto/Biography Studies* 32, no. 2 (2017): 279–282.

Huff, Cynthia, and Joel Haefner. "His Master's Voice: Animalographies, Life Writing, and the Posthuman." *Biography* 35, no. 1 (2012): 153–169.

Husserl, Edmund. *Ideas for a Pure Phenomenology and Phenomenological Philosophy. First Book: General Introduction to Pure Phenomenology*. Translated and edited by Daniel Dahlstrom. Indianapolis, IN: Hackett, 2014.

James, Erin. "What the Plant Says: Plant Narrators and the Ecosocial Imaginary." In *The Language of Plants: Science, Philosophy, Literature*, edited by Monica Gagliano, John C. Ryan, and Patricia Vieira, 253–272. Minneapolis, MN: U of Minnesota P, 2017.

Karban, Richard, and John Orrock. "A Judgment and Decision-Making Model for Plant Behavior." *Ecology* 99, no. 9 (2018): 1909–1919.

Kirksey, S. Eben, and Stefan Helmreich. "The Emergence of Multispecies Ethnography." *Cultural Anthropology* 25, no. 4 (2010): 545–576.

Krampen, Martin. "Phytosemiotics." In *Essential Readings in Biosemiotics: Anthology and Commentary*, edited by Donald Favareau, 257–278. Dordrecht: Springer, 2010.

Lansdown, Andrew. *Homecoming*. Fremantle, WA: Fremantle Arts Centre Press, 1979.

Leopold, Aldo. *A Sand County Almanac* and *Other Writings on Ecology and Conservation*. Edited by Curt Meine. New York, NY: Oxford UP, 1949.

Mabey, Richard. *The Cabaret of Plants: Botany and the Imagination*. London: Profile, 2015.

——. *Plants with a Purpose: A Guide to the Everyday Uses of Wild Plants*. London: William Collins, 1977.

Mancuso, Stefano. *The Revolutionary Genius of Plants: A New Understanding of Plant Intelligence and Behavior*. New York, NY: Atria, 2018.

Mancuso, Stefano, and Alessandra Viola. *Brilliant Green: The Surprising History and Science of Plant Intelligence*. Translated and edited by Joan Benham. Washington, DC: Island, 2015.

Marder, Michael. *The Philosopher's Plant: An Intellectual Herbarium*. New York, NY: Columbia UP, 2014.

——. *Plant-Thinking: A Philosophy of Vegetal Life*. New York, NY: Columbia UP, 2013.

Martin, Ann M. *A Dog's Life: The Autobiography of a Stray*. New York, NY: Scholastic, 2005.

Milkweed Editions. "Poetry: *Translations from Bark Beetle*." Accessed May 8, 2019. https://milkweed.org/book/translations-from-bark-beetle

Miller, Nancy K. "The Entangled Self: Genre Bondage in the Age of the Memoir." *PMLA* 122, no. 2 (2007): 537–548.

Rivas, Jesús, and Gordon Burghardt. "Crotalomorphism: A Metaphor for Understanding Anthropomorphism by Omission." In *The Cognitive Animal: Empirical and Theoretical Perspectives on Animal Cognition*, edited by Marc Bekoff, Colin Allen, and Gordon Burghardt, 9–17. Cambridge, MA: MIT P, 2002.

Russo, Linda. "Listening-Being: Some Unnamed Species of Porous Poems." *Jacket2*, June 14, 2015. https://jacket2.org/commentary/listening-being

Ryan, John C. "Gorge: Scriptorium | Tree | Excubitorium." Unpublished manuscript.

——. *Green Sense: The Aesthetics of Plants, Place, and Language*. Oxford: TrueHeart Academic, 2012.

——. *Plants in Contemporary Poetry: Ecocriticism and the Botanical Imagination*. New York, NY: Routledge, 2018.

——. "Poetry as Plant Script: Interspecies Dialogue and Poetic Collaboration in the Northern Tablelands Region of New South Wales." *Transformations* 30 (2017): 127–149.

Smith, Sidonie, and Julia Watson. *Reading Autobiography: A Guide for Interpreting Life Narratives*. 2nd ed. Minneapolis, MN: U of Minnesota P, 2010.

Solnit, Rebecca. "Acts of Hope: Challenging Empire on the World Stage." *Project SafeCom Inc.* Accessed May 8, 2019. https://www.safecom.org.au/acts-of-hope.htm

Tarlo, Harriet. "Introduction." In *The Ground Aslant: An Anthology of Radical Landscape Poetry*, edited by Harriet Tarlo, 7–18. Exeter: Shearsman, 2011.

Thoreau, Henry David. *Faith in a Seed: The Dispersion of Seeds and Other Late Natural History Writings*. Washington, DC: Island, 1993.

—— *Wild Fruits: Thoreau's Rediscovered Last Manuscript*. New York, NY: Norton, 2000.

Tompkins, Peter, and Christopher Bird. *The Secret Life of Plants: A Fascinating Account of the Physical, Emotional, and Spiritual Relations between Plants and Man*. New York, NY: Harper and Row, 1973.

Vieira, Patrícia. "*Phytographia*: Literature as Plant Writing." In *The Language of Plants: Science, Philosophy, Literature*, edited by Monica Gagliano, John C. Ryan, and Patricia Vieira, 215–233. Minneapolis, MN: U of Minnesota P, 2017.

Watts, Eric King. "Coda: Food, Future, Zombies." In *Voice and Environmental Communication*, edited by Jennifer Peeples and Stephen Depoe, 257–263. Houndmills: Palgrave Macmillan, 2014.

Weik von Mossner, Alexa. *Affective Ecologies: Empathy, Emotion, and Environmental Narrative*. Columbus, OH: Ohio State UP, 2017.

Wickens, Gerald. *Economic Botany: Principles and Practices*. Dordrecht: Kluwer Academic, 2001.

Wohlleben, Peter. *The Hidden Life of Trees: What They Feel, How They Communicate*. Carlton, VIC: Black, 2016.

"If a Tree Falls ... " : Posthuman Testimony in C. D. Wright's *Casting Deep Shade*

By Eamonn Connor

ABSTRACT

During a period marked by profound ecological transformations, there has been surprisingly little consideration of how testimony may operate as a mediating discourse between human and nonhuman. Based on a close reading of C. D. Wright's "memoir" *Casting Deep Shade*, this article reconsiders the subject positions of witnessing in posthuman terms.

> Witness tree, graffiti tree, tattoo tree, autograph tree, trysting tree, avenue tree, arborglyph, CMT (culturally modified tree), Presidents' Tree (for the one in Takoma Park carved with presidents from Washington to Lincoln in 1865; blown down in 1997). They say it really doesn't hurt the tree, all that carving. But harm and hurt are different. Beech bark is a tender thing.
> — C. D. Wright and Denny Moers, *Casting Deep Shade: An Amble Inscribed to Beech Trees and Co.*

Prior to C. D. Wright's unexpected death in 2016, she was engaged in years of research on the shared histories of humans and beech trees in the northeast US. Developing Wright's unique prosimetric style, the resulting book-length poem *Casting Deep Shade* is described by the author as a "memoir with beech trees," and the book's subtitle frames the text as a testimonial act: *An Amble Inscribed to Beech Trees and Co.*[1] "Amble" is how Wright describes her wanderings among the old-growth forests and estate arbors throughout the Ozark Mountains, a physiographic region in the US states of Missouri, Arkansas, Oklahoma, and Kansas. Yet, just as "ambling" denotes movement, "inscribed" remains static, denoting words written (or carved) on a surface, in witness.[2] Indeed, reviews have noted that *Casting Deep Shade* demonstrates "the power of poetry to conserve, preserve, and bear witness."[3]

Casting Deep Shade engages in poetic communication as a form of testimony, yet unlike Wright's earlier works it "affirms the need to witness" the world beyond the figure of the human.[4] Composed of ecological genealogies, colonial narratives, situated folklore, anecdotal asides, travel memoir, cultural commentary, and family autobiography, *Casting Deep Shade* begins as a passionate naturalist's ode to the beech tree and then branches out to touch on a dizzying array of topics, cultivating what Wright calls "an arboretum of the mind."[5] While commentators have noted the centrality of acts of witnessing to the memoir, Wright's elliptical prosimetric style challenges the singular "I" of the autobiographical pact and its investment in the human.[6] The figure of the subject who bears witness and provides testimony in *Casting Deep Shade* will thus be central to this article.

I want to suggest that Wright testifies from a subject position that exceeds her-*self*, producing a form of posthuman testimony from within an assemblage of human and nonhuman beings in the Ozark Mountains. Testimony here does not result from an encounter with the infinite alterity of the animal Other, but instead emerges from *inside* a community of beings that are entangled in specific and situated ways of living and dying. First, I will examine the posthuman position of the subject-witness who testifies in the memoir, leading me to suggest that the book itself is materially constructed as a type of "witness tree." I will then show how Wright seeks to produce, through various rhetorical gestures, a reader who is capable of bearing witness to testimony that emerges from beyond a bounded human subject. Since testimony is an epistemological act, I will also consider how testifying to becoming "beech-conscious" produces certain types of knowledge in the Ozarks, arguing that bearing witness to posthuman testimony produces an ethics based on *accountability* to "trees and tree-kin."[7]

A Subject-Witness in the Ozark Mountains

Casting Deep Shade is a three-panel hardcover tome that encloses a poem accompanied by striking black-and-white monoprints by photographer Denny Moers, hand-printed using analogue film and darkroom effects. In his introduction to the text, poet Ben Lerner describes the book as "a work in a perpetual state of becoming."[8] This is immediately acknowledged in the printing: the text bound to the center block (or trunk) is more embraced than contained by the wraparound cover, the pages (or leaves) just hanging on, at the threshold. The paratextual framing of the poem materializes Wright's later description of the beech tree: "Mono-layered leafers like the beech avoid blocking out each other's light by forming a jigsaw-like pattern to capture the light."[9]

When readers open the book, they encounter a color photograph on the inside panel. It is a portrait of sorts, although a far cry from conventional author photographs that work to essentialize the notion of selfhood and authorship. The image introduces Wright, standing bottom left of frame, as the "author" and subject-witness who provides testimony in the memoir. But she is not alone. At first glance, Wright appears to be standing beside a beech tree. On closer inspection, however, it becomes apparent that her feet are arched upon, "rooted with," to use Wright's phrasing, the "elephantine feet" of the tree.[10] Unlike traditional author photographs, Wright does not return the reader's gaze. Instead, we follow her gaze up to the ancient giant that exceeds the frame. This establishes that Wright is indeed the subject who bears witness and testifies, but she speaks within the memoir from a perspective *rooted* in a community of human and nonhuman beings, "Beech and Co."

In order better to understand this testimonial position, it is worth considering how Wright depicts the beech tree in *Casting Deep Shade*. She consistently seeks to focus our attention on the tree itself, not *the* beech tree in the figurative or abstract, but particular trees and groves as sites of assemblage. A beech tree is rooted in the earth, "[a]lways the risk of becoming rootbound"; trunk rising up, "tall trees … impressionistic creations"; branches splayed out, "whirling above the ground," with or without buds or leaves, "clover and climbing roses."[11] Is the beech tree, then, an object? If so, how should we define it? What is tree and what is not-tree—"Co." and "kin"? Where does *this* beech tree end and the rest of the Ozarks begin in *Casting Deep Shade*? Wright breaks off a piece of the tree and inspects it closely, "that watersmoothsilversatin bark."[12] She finds that it is populated by myriad organisms which have nestled beneath, "the scale (*C. fagisuga*) then the fungus (*Neonectria faginata* and *Neonectrica ditissima*)."[13] There are ladybugs and "parasitic wasps attacking aphids,"[14] and birds who make their nests and spread beech seeds so that the genus reproduces throughout the Ozark Mountains:

> Blue jay reflecting and scattering its blue, winging and cashing its nuts, foresting us with oaks and beeches (and in the case of the Steller's jays, pines). Bless their raucous, cheery souls. Jay, from Latin *gallus*, gay.
> Only 1 percent are egg robbers. *1%*.
> One was observed burying 2,000 beechnuts in one month.[15]

Wright's depiction of the beech tree establishes that it is not an object at all, but a situated gathering together of living beings and threads of relation in the Ozarks. *Casting Deep Shade* strives to build sites of attachment, connecting the living beings like the blue jays that bury beech nuts and congregate in and around the tree in a process which Wright describes as "*Inosculation*: intertwined and joined together."[16] This resonates strongly with Tim Ingold's argument that to inhabit the world is to "join in processes of formation."[17] The world that thus opens up to inhabitants is an environment without objects, populated by "things" that draw together lifeways: "The thing is a going on, or better, a place where several goings on become entwined. To observe a thing is not to be locked out but invited to the gathering."[18] Recall the first image we see of Wright on the inner panel of the book—she gazes up at the beech tree, observing a "thing" rather than an "object," and as we shall see, these acts of witnessing increasingly embed her within the "gathering" over the course of the memoir. The "thing" is Ingold's formulation of an assemblage, a concept to which ecologists have turned to avoid "the sometimes fixed and bounded connotations of ecological community."[19] Wright depicts the beech tree as an assemblage, an open-ended gathering in which things *become* together, for, as Anna Tsing argues, "[a]ssemblages don't just gather lifeways, they make them."[20] The subject-witness who provides testimony in *Casting Deep Shade* is a subject in a perpetual state of becoming. The origin of the testimony is not a complete, bounded subject, and testimony emerges from within the assemblage of Ozark beech groves, an assemblage that is never static but always *in process*—a becoming "beech-conscious."[21] Wright thus disturbs the principle that tethers the narration of "lives" to the singular and anthropocentric embodiment of "a life."[22] In *Casting Deep Shade*, the embodied and embedded human being represented in autobiographical narrative becomes part of a posthuman assemblage of bodies, history, and matter in the Ozark Mountains.

Witness Trees and Co

> The beeches of Oradell, NJ, are said to have witnessed certain episodes of the American Revolutionary War. Many trees went down, but the beeches survived the nor'easter of 2010, Hurricane Irene, and Superstorm Sandy.
> — C. D. Wright and Denny Moers, *Casting Deep Shade: An Amble Inscribed to Beech Trees and Co.*

Recognizing that Wright testifies from a subject position beyond her-self has significant ramifications for how we understand the text as a testimonial document. Testimony is a mediating discourse, which is to say that the past and present are mediated through testimony, and the production of testimony functions as an intermediary between individuals, communities, and the social field, and, in this case, between human and nonhuman. *Casting Deep Shade* textually mediates Wright's acts of witnessing to an addressee, an audience-witness, who is capable of bearing witness to posthuman testimony without rejecting or transforming it. In order better to understand this process of mediation, we can turn to Wright's formulation of the "witness tree," a figure that recurs frequently throughout the book. In *Indian Trails of the Warrior Mountains*, a field guide to Native American history referred to within *Casting Deep Shade*, a witness tree,

1. Bears manmade markings or evidence of non-naturally occurring reformative growth patterns.
2. Serves as a signpost or communication device.
3. Acts as a message board, boundary marker, warning sign, or has evidence of domestic usage.
4. Is a repository of vital information or record keeping.
5. Contains information pertaining to any cultural or historical event which occurred during the lifespan of the tree.[23]

I want to suggest that *Casting Deep Shade* fulfils these criteria and the book itself is constructed as a type of "witness tree." It bears the markings of tree lore ("Beech in the house interferes with a spirit's passing"[24]) and indigenous and colonial histories ("Cherokee in presyllabaric code"[25]); it communicates through a foliate language ("The hanger-on-until-spring leaf is *marcescent*"[26]); it is a repository of vital information about the genus of trees (beech) and its family (*Fagaceae*) and environmental history ("In the Pleistocene, beech spanned the continent"[27]); and it testifies to cultural and historical events ("The beeches of Oradell were scheduled to bear silent, unsworn witness to Benedict Arnold"[28]). *Casting Deep Shade* is a repository of record-keeping, full of esteem for arborists, indigenous knowledges, and folklore, for they are—like Wright—committed to keeping language and landscape particular, unpredictable, "testimonial."[29] It is a message board and a boundary marker: "The beech has to do with thresholds. Used to mark boundaries."[30] Yet, these "boundaries" are frequently shown to be fluid, dependent on the material conditions of assemblages rather than serving to separate or categorize living beings or cultural histories and concepts. Like the "witness tree," *Casting Deep Shade* is a site of inscription, a kind of "slow-growth kinetic sculpture" that proceeds at an amble to "preserve in living shape certain particular treasures of the past and expectations for the future."[31]

The poem is also, very literally, an arborglyph inscribed on pages that began as tree flesh. It is materially composed of beech woodgrain paper, and this materiality is foregrounded when Wright urges the reader to "read the leaves" closely.[32] She speaks of writing the poem by "using a beech branch as a pen. … it will enable a bok," and thus situates the origins of art in environment: "Beech is Anglo-Saxon *bok*: book, document, or charter; and *beche* in Middle English, *bech, beetch, beech* from Old English; *bece* of Germanic origin, *buoh* in Old High German, *boke* in Middle Low German, *baike* in Modern Low German, *in modern* German

Buche, Old Norse *bok*, and Dutch *boek* and Danish *bog* and Swedish *bok*, all meaning both book and beech."[33] Wright makes it clear that the book was formed by joining with and following forces and flows of materials in the Ozarks. In doing so, she does not seek to replicate static forms in the world. As the plant grows from its seed, *Casting Deep Shade* grows from acts of witnessing that articulate relations in motion. Karen Barad's theory of "agential realism" continues a tenet of post-structuralism in critiquing representationalist theories for upholding a gap between thoughts and the world, and for conceptualizing the project of epistemology as the pursuit of accuracy between the two.[34] Representationalism assumes "that which is represented is held to be independent of all practices of representing."[35] *Casting Deep Shade* does not attempt to mirror preexisting phenomena to organize its account of the natural and social world of the Ozarks. When Wright writes "Every time the tree works the leaves dream," she is playing with the double meaning of "leaves," referring to not only the literal leaves of a beech tree, but also the verb to "leaf through" the fleshy pages of the book the reader holds in their hands.[36] By making it clear that *Casting Deep Shade* is, very literally, a *bok*, Wright frames the poem as a type of embodied inscription: "The media … is living."[37] The "thing" represented is not held to be independent of her embodied poetic practice of "ambling." She calls the poem an "improvisation," and this entails following the ways and relations of the Ozarks as they become, rather than connecting up in reverse a series of points already traversed.[38] Similarly, Moer's photographs throughout the book do not reproduce preconceived ideas or figures of trees but work with chemicals and materials from forests in the Ozarks to "transform from the literal to the imagined via accidental contingent design."[39]

These processes of production and inscription suggest that *Casting Deep Shade* is a type of "witness tree": materially composed of tender bark, bearing arborglyphs engraved by a "branch," testifying through the markings and evidence of the entangled social, cultural, and environmental histories of the Ozarks. It is a communication device, a source of vital information and record-keeping, and a facsimile of a tree's growth outwards—a cumulative chronology in rings of testimony. The book is a gathering together of threads of life in the Ozarks, "trees and tree-kin" and "Beech and Co.," and it thus mediates a posthuman testimony.[40] *Casting Deep Shade* is the "witness tree" upon which Wright has inscribed her acts of witnessing as a relational subject.

Bearing Witness to *Casting Deep Shade*

Having established a subject-witness within an assemblage, whose testimony is mediated by *Casting Deep Shade* as a type of witness tree, what is the location of the audience-witness here? Dori Laub has attested to the importance of being an "authentic witness" who can recognize the truth of the experience being testified to.[41] Without this aspect of testimony, Laub argues, there is a "collapse of witnessing," whereby the testimonial account is not recognized and fails to produce knowledge.[42] *Casting Deep Shade* works to produce a reader who is capable of bearing witness to the posthuman testimony that Wright delivers as a subject-witness inside the assemblage of "tree and tree kin" in the Ozarks.[43] She actively points to a material-discursive reading practice: "The mono-focal experience of the *bok* is the heart and soul of what it means to read."[44] The use of the term "*bok*" highlights the origins of books in beech, and this suggests that reading *Casting Deep Shade* is a type of material experience with the beech tree, rather than a passive encounter with a testimonial document whose words "represent" the lifeways of the tree. When writing about climate crisis, for instance, Wright plays with the material relation between the book and the beech: "Knowing the general outline of the argument of the Holocene extinction put forth by Elizabeth Kolbert in *The Sixth Extinction: An Unnatural History*, I have not yet brought myself to the pain of cracking its spine, to be taken to the brink."[45] Wright associates the pain of encountering climate-crisis testimony in Kolbert's book with "the pain of cracking its spine." The "mono-focal experience" that defines beech-consciousness is thus an *embodied* ethical position oriented toward specific relations and materialities in the Ozarks. It is an ethics rooted in this place, these trees, these people. *Mono* denotes singularity, and this singularity is materially registered in the paratextual framing of the book. After "cracking its spine," the book unfolds into an expansive three-page folder. Due to its size and the fragile connections between the panels, it cannot be read while reclining on a couch or in bed. As it lacks a binding stem, the reader must always be touching the book; otherwise, the poem will snap shut and the testimonial encounter will end. Lest we forget the delicate material relation between book and *bok*, Wright reminds us that "[b]eech bark is a tender thing."[46] This suggests that the testimony offered by *Casting Deep Shade* is textually fragile. The care that must be taken while reading should be understood as connected to Wright's stated desire to produce "sustained acts of attention, slow exposures of gaze" in the reader.[47] Bearing witness here is an ethical act that attends to singularity, the "mono-focal," material experience of encountering testimony that emerges from beyond the human subject.

The architecture of the poem also works to produce a certain "mono-focal" experience of reading that does not operate through narrative accumulation but sensorial encounters. Brought before *Casting Deep Shade*, we must abandon our accustomed ways of reading, for, as Lerner notes, "this is also a book about looking."[48] We do not scan for information, the way we might do when we scroll through media on a screen,[49] and there is no strong narrative compulsion to the book-length poem. The compressed lines and the silence that surges from the margins and between polyphonic fragments make the gaze linger and alight. Passages distort time by intensifying the senses, inviting the reader to bring to mind the taste and texture of "horse chestnuts aplenty … rich in oil, tanning."[50] The poem's fragments resist paraphrase, which is to say that the poem is not particularly invested in producing knowledge *about* the beech tree or the speaking "I" of the memoir. There are occasional didactic asides: "Beech, *Fagus*. Family Fagaceae. Alternating leaves of the *sylvatica* (European) crenated, of the *grandifolia* (American) crenulated; the former a little wavy, the latter a little toothy."[51] However, these blocks of information, composed of etymologies and biological science, are frequently interrupted by poetic interludes, like "the former a little wavy, the latter a little toothy." These effusive, imaginative disruptions frame scientific discourse as simply one thread in the deep wells of knowledge that inosculate in the Ozarks. *Casting Deep Shade* performatively enacts a type of testimonial "dispersal"—tracing the disparate threads of relation within the assemblage. Wright writes the following about beech-tree propagation:

> Dispersed by wind. Says one.
> Not dispersed by wind. Says another.
> Dispersed by wildlife. Says one.
> Dispersed by blue jays. Says another.
> Not dispersed by jays. Says one.[52]

A "dispersal" is above all an act of distribution, and by rhetorically enacting the dispersals that mark the assemblage, Wright gives shape to witnesses as distributions or assemblages of human and nonhuman beings. In the passage above, the "jays" are a crucial actant, among many others, in propagating the threads of life that congregate in and around the beech tree, and determine the conditions of possibility for the assemblage itself. The frequent interruptions and "dispersals" awaken the reader's concentration—the shifts in perspective and register require a focused reading that attends to the flights of becoming within the assemblage, that resists the rush to unity for the complexity of entanglements. When we read *Casting Deep Shade*, our eyes may wander down the pages, but they also linger on charged fragments: we move forward, we skim back, we examine closely, we *crack* the spine of the book in our hands. Wright's poetry produces a kind of "fluid but deep concentration," retraining our minds to dwell in the open awareness that marks beech-consciousness.[53] When we amble through the poem, our typical capacities of *making sense* are destabilized and we must feel our way forward with our unknowing. I am given fleeting glimpses of what pulses beneath the words: the mysteries of relationality in the Ozarks in which no-thing and no-one are truly separate. Reading *Casting Deep Shade* attentively means tracing the situated material interchanges across human bodies, animal bodies, and tree bodies. "Some of us," Wright says, "do not write particularly for pleasure or instruction, but to be changed, charged."[54] To crack the spine and "be taken to the brink," to read to be changed, requires a witness that can be attentive to the flux of becoming in beech groves.

In order to understand why *attention* is necessary for a reading that is capable of bearing witness to the testimony of *Casting Deep Shade*, it is important to understand how the book testifies to a duration. Just as the "author" is not a bounded individual, the book does not testify to single, delimited events. Jacques Derrida notes that the assumption in testimony is that we testify to an "instant."[55] For example, in a legal context, the witness testifies, in an instant, to their presence at a prior instant. In contrast, Wright does not simply attempt to *capture* the reader's attention to bear witness to an "instant," but works to produce sustained acts of attention that can bear witness to the duration of *becoming* "beech-conscious." It is worth turning for a moment to the figure of the "adequate witness" in the work of Leigh Gilmore. In *Tainted Witness*, Gilmore takes the unequal role of judgment in bearing witness as a starting point to discuss the myriad ways in which women's testimonial accounts are discredited. She searches for an "adequate witness," a "moving target" who receives testimony without transforming it by "suspicion."[56] Albeit in a very different context, I want to suggest that, in performatively enacting the "dispersals" that mark witnessing from inside the assemblage, Wright aims to produce a kind of attention in the reader that attends to her testimonial accounts of becoming within the Ozarks, while resisting the rush to judgment or transforming the testimony into a representation that functions didactically as a "field guide" or merely stands in symbolically or metaphorically for something else. *Casting Deep Shade* appeals for an "adequate witness" who is capable of "taking in a witness tree and its ramifications" while avoiding affective identification practices like empathy and sympathy, which extend ethical rights to nonhumans on the basis of their similarities to us.[57]

Becoming "Beech-Conscious"

A scheme emerges here: the subject-witness is Wright, testifying from a subject position beyond her-self within the assemblage of "Beech and Co." in the Ozarks. This testimony is mediated by the physical text, inscribed on a "witness tree," which actively appeals to an "adequate witness" who is capable of bearing witness to a duration—a becoming "beech-conscious." Since testimony is an epistemological act, this leads us to the question: What knowledge is produced in the testimonial transaction?

The poem begins with descriptions of the beech tree taken from the National Forest Register: "Poses no significant litter problem | Ranks as 'not particularly outstanding' according to the Forest Service."[58] The opening pages provide a broad genealogy of human-beech relations: "Iron Age man made beechnut flour | Native Americans made beechnut flour."[59] Wright utilizes an emotionally detached academic discourse: "The pollen record keeps going back and back | Pollen from pre-Roman peats has been found in the UK."[60] After these didactic statements, indented by line breaks, she begins to introduce blocks of text that provide genetic and species information, and Latin etymologies from texts like the *Manual of Woody Plants* and *Portraits of Forest Trees*.[61] Slowly, she begins to interweave excerpts of Romantic poetry, from Robert Frost and John Milton, which describe beech trees: "Thick as autumnal leaves that strow the brooks."[62] However, this line from Milton is directly juxtaposed with a previous line from the *Manual of Woody Plants*, which notes, "In autumn … the leaves still cling to their stems."[63] We learn that, since it was a beech forest surrounding the ancient Benedictine monastery at Vallombrosa where Milton claimed to be staying when writing his poem, the poet either never visited or willfully distorted his surroundings when he wrote the poem. This is one of several occasions in the text where Wright playfully mocks canonical writers like Milton, Virgil, John Keats, Alfred, Lord Tennyson, Mark Twain, Henry David Thoreau, and Voltaire for misrecognizing the beech tree in their writing.[64] She shows that they consistently misname or ignore the materiality of the trees, even as they instrumentalize them to reify an essentially humanist and anthropocentric Romantic worldview. She chastises them for their "lack of attention" to the trees that they employ as symbols and metaphors.[65] In showing how these poets lack beech-consciousness, she establishes the first elementary characteristic of becoming within a posthuman assemblage of "tree and tree-kin"—namely, the human must be attentive to the materialities of the living beings around them. This includes noticing that beech leaves cling to their stems in autumn and knowing the difference between an oak, a chestnut, and a chinkapin.[66] Part of the "mono-focal" experience of the beech is recognizing that a tree is a singular living being.

Wright compares the figurative metaphors of the poets to the writings of prominent arborists who produce natural histories of the Ozarks: "All of Larry Lowman's information about beech aspect and behavior is borne out in the literature again and again."[67] However, she makes it clear that this information is only meaningful insofar as it allows for a recognition that facilitates a material encounter: "but there is nothing quite like a close encounter of the other kind, either at the elephant foot of one of the elder survivors of human disturbance and vengeful weather or from the mouth of a native plant master."[68] She testifies that there are knowledges that escape the humanist representationalism of the poets, and the discourse of the scientists and arborists, which are produced through direct material encounters with beech trees and the testimonies of the indigenous peoples of the Ozarks.

Wright suddenly departs from the linguistic pattern of block text and inserts indented and bracketed lines that give a clearer sense of the sort of knowledge which interests her: "If you take root, you'll go deep. | (Always the risk of becoming rootbound)."[69] This break in the established pattern suggests that this knowledge emerges in the cracks, is "rooted" in the spaces between the intersection of human and environmental histories. She follows these lines by noting that beech trees are not "rooted" in the way we might think they are—"A taproot is not the securing component of the beech, which instead spreads its elephantine feet out along the surface"[70]—and claims that to "be rooted" is the most important and least recognized need of the human—"A human being has roots by virtue of his real, active and natural participation in the life of a community, which preserves in living shape certain particular treasures of the past and certain particular expectations for the future."[71] She then indents another stand-alone line: "Minus the expectations, trees and humans do manifest a common gestalt."[72] The text is accompanied by a color photograph, taken by Wright on one of her visits to beech groves in the Ozarks, of the "elephantine feet" of a beech tree spreading across the grounds of a park.[73] This makes visible her claim that beech trees are not "rooted" in the way we may think they are—that is, via underground networks invisible to the human. This testimony produces a specific knowledge: trees and humans are *both* rooted by virtue of their active and material participation in the life of a shared community. However, Wright does not write that trees and humans "share" a common gestalt; she is unequivocal—they "manifest" a common gestalt. This tells us that while humans in the Ozarks are always already embedded in a community of (non)human beings, to be "rooted" (the most important need of the human) requires a "real" and "active" participation in the life of that community. Being "rooted" is not a passive inclusion. Rather, the act of bearing witness and testifying from within the assemblage, and adequately bearing witness to that testimony, necessitates material encounters. Wright is testifying here to the emergence of beech-consciousness—the recognition of herself and the world from the perspective of a posthuman assemblage in the Ozarks means relinquishing one's figurative relations with beech trees and actively recognizing and engaging in the material relations that constitute a situated community.

On the following page, Wright introduces the first-person "I" for the first time in the poem, as she recalls a trip she had taken as a child to the "hill-country splendor" of Missouri.[74] She writes: "I was not beech-conscious at the time."[75] She testifies that, at this point in her life, she was only "semi-aware" of the groves.[76] She lacked the awareness to recognize that she was a "by-product of the Ozark Mountains."[77] She was not yet able to recognize the beech trees and hardwoods as "her standing brothers and sisters."[78] So, becoming "beech-conscious" is a move from semi-awareness to a complete awareness of the way in which we are ecologically embedded in local communities. It means recognizing the ways in which we are all "by-products" of specific, situated ecologies, embedded in relations of kinship that exceed human communities. This awareness does not happen at any one instant, but is rather a *process* made possible through material encounters. *Casting Deep Shade* does not function as a step-by-step guide or manual to achieving beech-consciousness. Instead, Wright plots her "becoming" as a series of linked instants and, in doing so, testifies to a duration: beech-consciousness is always *emergent.*

At this point in the memoir, much like the Romantic poets, Wright consistently misrecognizes trees: "I curse-swished Spaldings at age 10 … or was it a pawpaw tree."[79] While reflecting on her youth, she describes her nearsightedness as a "minor tree-related manner."[80] How exactly could it be a "tree-related manner"? Wright recalls how the only trees that her parents allowed her to climb as a child were trees which had been felled. She encountered only dogwoods, "squat, flouncy things," and she describes how she thought they were "on my level."[81] Her nearsightedness meant that standing, living trees were naught but "impressionistic creations" to her eyes—they only "individuated" upon her "climbing to their level."[82] As she testifies to the moment of that realization, she indents a bracketed line: "(The buds of tree consciousness.)"[83] Here, she connects to the slow climb "to their level" as the beginnings, the "buds," of becoming "beech-conscious." Again, she is testifying to the idea that becoming "rooted" in the community of "Beech and Co." is made possible by active material encounters with the nonhuman members of said community. She juxtaposes this testimony with the story of a nearby house in Arkansas named Twelve Oaks: "a sternly aristocratic, white stucco house on Hwy & en route to the Buffalo River … The drive lined with a dozen imposing oaks. A tornado took each one of them out like soda straws. Without them the house looked like an asylum. Exactly like an asylum."[84] This passage functions as a sort of cautionary tale. The budding beech-consciousness to which she is testifying contrasts vividly with the grim image of an austere house stripped of its trees. The house becomes "[e]xactly like an asylum," associating a lack of beech-consciousness with disconnection and madness, and pointing the reader toward the consequences of environmental destruction. For Wright, it is becoming *with* those "slow-growth kinetic sculptures" that "tethers life."[85] She thus suggests that people in the Ozarks who fail to recognize their belonging within a community of human and nonhuman beings, who fail actively and materially to manifest that belonging, that common gestalt, become untethered from the threads of life which make living and dying possible in that community. This resonates with the argument made by Cary Wolfe that we are "not that auto- of autobiography that humanism gives to itself."[86] Wright testifies that there is a certain confinement and madness to the isolated and separate humanist subject, and connecting imagery of "soda straws," preeminent symbols of twenty-first-century pollution, with the violence of a tornado provides a grim vision of a future that untethers the human from communal life.

Becoming "beech-conscious" is also a process that illuminates the entanglement of environmental and colonial histories in the Ozarks. For instance, "witness trees" throughout the groves bear cultural histories through the presence of Native American arborglyphs: "Regarding the witness tree: Beech is preferred for carving in the East, aspen in the West."[87] The "witness trees" in the East, throughout the Ozark Mountains, are beech precisely because of the colonial exploitation of the indigenous Cherokee populations: "The color for the Long Hair Clan of Cherokee (also known as Twister, Wind, or Hanging Down Clan) is yellow. Their tree is beech."[88] Wright notes that there must be many Cherokee arborglyphs that still endure, but many people in the Ozarks would be unaware of their presence, since the lettering would be far from the ground and much expanded with the growth of the trees. This recalls her earlier description of lacking beech-consciousness because she could not and would not meet the trees "on their level."[89] By testifying that, in the Ozarks, environmental history and exploitation are inextricable from colonialism and cultural erasure, *Casting Deep Shade* shows that becoming "beech-conscious" necessarily involves becoming aware of the exploitation of indigenous peoples and the complex entangled histories within a situated community. Any reading or encounter with the beech tree, our "standing brothers and sisters,"[90] which fails to produce that knowledge is an inadequate act of witnessing and diametrically opposed to beech-consciousness.

Later in the poem, Wright tells the story of meeting a physically vulnerable English poet in Arkansas with a rare, agonizing disease: "When drugs offered no relief or necessitated tapering off, NATURE, she vowed (in all caps), was the only healer."[91] Afflicted since birth, the poet recalls suffering greatly one day as a child, going outside and lying on her back in the grass beneath a beech tree: "When she stood up she beheld a glimmer of blue silhouetting her body that quickly dematerialized."[92] While the phenomenon never recurred, leaning on trees or lying on the earth continued to soothe her. Crucially, "physical contact was essential to receive the succor offered."[93] Wright testifies to the idea that there must be some material contact for beings, human and nonhuman, to "succor," to support each other within the Ozarks: "Embarrassed to hug it [the tree] then and there, but tentatively, I did. Who knows when an otherwise inexplicable transference of strength might strike."[94] Strength is passed, distributed, and dispersed throughout the members of the community. However, while the beech tree is shown to offer "succor," the human is frequently shown by Wright to fail to reciprocate. On the following page, Wright juxtaposes the story of the vulnerable English poet with a retelling of the story of Joan of Arc. She notes that the beech tree was

> Never a hanging tree:
>
> but Joan of Arc was thought by her examiners to be receiving satanic messages from a Fairy Tree, a beech, alas.[95]

Wright's retelling of the execution of Joan of Arc testifies that human beings have long been aware of the presence of beech trees, and aware of their capacity to communicate and change the outcome of events, but framed this communication as mystical and dangerous, associating it with a threatening, nonconforming feminine. Joan is burned alive in Wright's account because she "heard the (demonic) tree speak."[96] Wright connects this account of Joan with the "assassins" of the beech tree, *Phytophthora*, a pathogen that has infected 40% of beeches in the northeastern US, attracting ambrosia beetles and chestnut borers, which "set upon the stressed."[97] The wilted foliage makes the European beech subject to burning and relational to Joan in its precarity: "The Europeans burn easily. Once the foliage is stripped they are subject to sunscald."[98] It is here that Wright begins to consider the ethical dimensions of beech-consciousness, as she explicitly claims that both humans and trees in the Ozarks share the US Department of Agriculture (USDA) hardiness zones 4–7, or 3–8, "depending on the literature."[99] A "hardiness zone" is a geographic area defined to encompass a range of climatic conditions relevant to plant growth and survival. While it explicitly refers to flora, Wright testifies that the USDA zones are in fact the climatic conditions relevant to the growth and survival of *both* human and nonhuman beings in the Ozarks.

By foregrounding the material inter- and intra-connections across bodies in the Ozarks, the testimony that emerges from *Casting Deep*

Shade denies anthropocentric exceptionalism by considering all species as intermeshed within larger currents. If, as Wright testifies, "the grey-water days are upon us,"[100] then she is clear that the only appropriate ethics involves the "pain of cracking [the] spine," a spine that would separate human, trees, and other Earth "others." The ethical knowledge produced here is not grounded in a situated response to a radically exterior|ized other, but about accountability for the lively relationalities in which we are embedded in specific environments—in this case, USDA zones 4–7 or 3–8 in the Ozarks. This recalls the situated ethics espoused by Donna Haraway, who writes, "Nobody lives everywhere; everybody lives somewhere. Nothing is connected to everything; everything is connected to something."[101] Wright testifies that beech-consciousness makes the question of accountability more insistent, as the intra-active production within assemblages means that we actively contribute in bringing forth the world in its specificity, ourselves included, "split-open ribcages, palpitating hearts in hands."[102]

As Wright, the subject-witness of the testimonial chain, becomes increasingly "beech-conscious," she begins to rematerialize her language. Scott Knickerbocker has described this process, of cultivating an active and productive imagination concerning the "language" of nonhumans, as "sensuous poiesis"—"the process of rematerialising language specifically as a response to nonhuman nature."[103] Wright's poetry and prose change as she becomes increasingly conscious of the materialities of the assemblage and seeks actively to respond to the textures and sounds that she encounters. She writes: "The watersmoothsilversatin bark is an adaptation from its tropical beginnings, fending off the ephiphytic plants."[104] Humans and trees become "inosculated"; swaying funnel-shaped treetops are described as "infunibular"; smooth, hairless leaves are "glabrous"; and ancient beeches waiting to be felled are "grandegrannydames."[105] Portmanteaus like "watersmoothsilversatin" and "grandegrannydames" are utilized for their intensely onomatopoetic effects, and these descriptions of the beech trees complement the textural photographs that adorn the pages opposite her writing. The "buds" of tree consciousness infuse language through the double meaning of words like "sappily," "brood," "root," "heartwood," "harvest," "branch," "weeping," and "bark," and death is described as the "long-term dirt nap."[106] Wright is explicitly aware of the interplay between language and nature, book and *bok*:

First page
of literature
in Sanskrit
on beech
the runic tablets
on beech

Firstbooks [Database][Mismatch]
were beech
in Sanskrit
the Vedas
Who know
who wrote
Old English
on bound
beech
bark.[107]

In making clear that it was always our relations with beech trees, among innumerable other nonhuman beings, that made human communication and art possible, Wright stresses the depths of our ethical accountability. We are at stake. Wright does not observe the communities of beech groves from some outside position; she testifies that she is part of the Ozarks in its ongoing emergence, and that this has ethical ramifications. Near the end of the poem, she begins employing complex compound sentences, spanning, at times, a couple of hundred words, which innovatively imitate the movements of the beech tree, not a static object, but a "thing" that gathers together threads of life: "When one experiences a grove of beeches, it is a life force revealed."[108] She goes on to suggest that one of the ways that we can respond to our ethical *accountability* is to join with the forces and flows within the assemblage. When we bear "adequate witness" to beech trees, we can develop innovative and unanticipated responses to the environmental crises that we are facing. For instance, Wright describes the way that Japanese scientist Koryo Miura recently conceived of a method to fold an array of solar panels by mimicking beech leaves unfolding from the bud: "Further applications include a subway map and heart stents."[109] Underneath this passage, she includes a panorama of Miura's solar-panel designs and the accordion folds of a beech leaf. The image is recalled again at the very end of the poem. *Casting Deep Shade* becomes increasingly effusive, until Wright suddenly breaks from the complex compound sentences and reproduces the beautiful silent poem "Fisches Nachtgesang" ("Fish's Night Song"), by Christian Morgenstern.

"Fisches Nachtgesang" consists only of patterns of dashes and brackets printed to replicate scales, waves, and bubbles. It is an example of concrete poetry, yet its visual shape is abstract. It is an aural poem, marked by strokes and circular arcs that graphically describe the accented and unstressed syllables of a poem. Yet it evokes no discernible sounds. It is a wordless testimony. The lines are not connected to form a solid figure, and the shape of the fish remains just out of view. It is up to the reader, the audience-witness, to fill in the blanks, to give form to the shape. By refusing to provide any explanatory, didactic information about the origin of the poem, Wright evokes the earlier image of Miura's solar panels and the ridged accordion folds of the beech leaf. Underneath this visual poem, Wright includes a single line: "Strange otherworld. But it is inside this one, just as Éluard said."[110] She challenges the reader to perceive this "otherworld … inside this one"—the more-than-human world that can only be perceived, the text suggests, through a process of becoming tree-conscious. Wright thus gives priority to processes of formation over their final products, and to material flows and transformations over states of matter. Form, to recall the words of Paul Klee, is death; form-giving is life.[111] "Fisches Nachtgesang" bespeaks possibility, a possibility to which we are frequently blind because we lack the proper consciousness. Wright testifies that this is possible under certain material conditions. In the Ozarks, this is a process of becoming "*beech*-conscious," which illuminates the entangled social, cultural, and environmental histories unique to that region and community. Wright witnesses from inside the assemblage, testifying that humans and trees in the Ozarks are part of "the same mealy, meaty family."[112] In *Casting Deep Shade*, material encounters with "witness trees" and "tree-kin" leave Wright "changed, charged." She testifies to this change, a becoming "beech-conscious," and seeks to produce a certain type of reading that might leave the audience-witness of the text "charged" and "changed" in their encounter with a "witness tree." To bear "adequate witness" to the testimony that emerges in *Casting Deep Shade* means to resist the rush to judgment and open oneself to the pain of "cracking its spine," in the knowledge that the encounter with testimony could "trigger a reset of your entire cosmology."[113]

When I close the final "leaves," *Casting Deep Shade* does not free me from its grip. The force it exerts *charges* my consciousness. I close the spine with a sense that I have been complicit in preventing my capacity to relate. Wright makes it clear that life is open-ended: its impulse "is not to reach a terminus but to keep on going."[114] She testifies to a renewed sense of accountability, to the ongoingness of life beyond the human subject, a "life" in which we are embedded and co-constituted, and for which we are accountable. In a final act of generosity, a final testimony to the fact that no subject is ever truly bounded or alone, Wright chooses to end her memoir with a line from her late friend, the poet and environmentalist W. S. Merwin: "On the last day of the world | I would want to plant a tree."[115]

University of Glasgow

Notes

1. Wright and Moers, *Casting Deep Shade*, 1.
2. Muske-Dukes, "Poet."

3. "Casting Deep Shade."
4. Muske-Dukes, "Poet."
5. Wright and Moers, *Casting Deep Shade*, 78.
6. Muske-Dukes, "Poet." See also Fishman, "Poetry Review."
7. Wright and Moers, *Casting Deep Shade*, 23.
8. Wright and Moers, *Casting Deep Shade*, 23.
9. Wright and Moers, *Casting Deep Shade*, 65.
10. Wright and Moers, *Casting Deep Shade*, 65, 21.
11. Wright and Moers, *Casting Deep Shade*, 21, 31, 31, 64.
12. Wright and Moers, *Casting Deep Shade*, 203.
13. Wright and Moers, *Casting Deep Shade*, 131.
14. Wright and Moers, *Casting Deep Shade*, 129.
15. Wright and Moers, *Casting Deep Shade*, 209.
16. Wright and Moers, *Casting Deep Shade*, 207.
17. Ingold, "Bringing Things to Life," 3.
18. Ingold, "Bringing Things to Life," 3.
19. Tsing, *Mushroom*, 23.
20. Tsing, *Mushroom*, 23.
21. Wright and Moers, *Casting Deep Shade*, 24.
22. Whitlock, "Post-ing Lives."
23. Marshall and Walker. *Indian Trails*, 18.
24. Wright and Moers, *Casting Deep Shade*, 69.
25. Wright and Moers, *Casting Deep Shade*, 65.
26. Wright and Moers, *Casting Deep Shade*, 64.
27. Wright and Moers, *Casting Deep Shade*, 6.
28. Wright and Moers, *Casting Deep Shade*, 93.
29. Wright and Moers, *Casting Deep Shade*, x.
30. Wright and Moers, *Casting Deep Shade*, 212.
31. Wright and Moers, *Casting Deep Shade*, 22.
32. Wright and Moers, *Casting Deep Shade*, 12.
33. Wright and Moers, *Casting Deep Shade*, 69, 50.
34. Barad, *Meeting the Universe Halfway*, 46.
35. Barad, *Meeting the Universe Halfway*, 46.
36. Wright and Moers, *Casting Deep Shade*, 125.
37. Wright and Moers, *Casting Deep Shade*, 164.
38. Wright and Moers, *Casting Deep Shade*, 33.
39. Wright and Moers, *Casting Deep Shade*, 40.
40. Wright and Moers, *Casting Deep Shade*, 23, 1.
41. Felman and Laub, *Testimony*, 80.
42. Felman and Laub, *Testimony*, 80.
43. Wright and Moers, *Casting Deep Shade*, 23.
44. Wright and Moers, *Casting Deep Shade*, 125.
45. Wright and Moers, *Casting Deep Shade*, 63.
46. Wright and Moers, *Casting Deep Shade*, 62.
47. Wright and Moers, *Casting Deep Shade*, xi.
48. Wright and Moers, *Casting Deep Shade*, xi.
49. Xie, "Restoration," 2.
50. Wright and Moers, *Casting Deep Shade*, 26.
51. Wright and Moers, *Casting Deep Shade*, 4.
52. Wright and Moers, *Casting Deep Shade*, 209.

53. Xie, "Restoration," 2.
54. Wright and Moers, *Casting Deep Shade*, 55.
55. Derrida, "Demeure," 36.
56. Gilmore, *Tainted Witness*, 2.
57. Wright and Moers, *Casting Deep Shade*, 24.
58. Wright and Moers, *Casting Deep Shade*, 2.
59. Wright and Moers, *Casting Deep Shade*, 2.
60. Wright and Moers, *Casting Deep Shade*, 3.
61. Wright and Moers, *Casting Deep Shade*, 8–9.
62. Wright and Moers, *Casting Deep Shade*, 20.
63. Wright and Moers, *Casting Deep Shade*, 20.
64. Wright and Moers, *Casting Deep Shade*, 20, 56, 59, 63, 108, 151, 239.
65. Wright and Moers, *Casting Deep Shade*, xi.
66. Wright and Moers, *Casting Deep Shade*, 25.
67. Wright and Moers, *Casting Deep Shade*, 21.
68. Wright and Moers, *Casting Deep Shade*, 21.
69. Wright and Moers, *Casting Deep Shade*, 21.
70. Wright and Moers, *Casting Deep Shade*, 21.
71. Wright and Moers, *Casting Deep Shade*, 22.
72. Wright and Moers, *Casting Deep Shade*, 22.
73. Wright and Moers, *Casting Deep Shade*, 21.
74. Wright and Moers, *Casting Deep Shade*, 24.
75. Wright and Moers, *Casting Deep Shade*, 24.
76. Wright and Moers, *Casting Deep Shade*, 24.
77. Wright and Moers, *Casting Deep Shade*, 25.
78. Wright and Moers, *Casting Deep Shade*, 25.
79. Wright and Moers, *Casting Deep Shade*, 26–27.
80. Wright and Moers, *Casting Deep Shade*, 31.
81. Wright and Moers, *Casting Deep Shade*, 31.
82. Wright and Moers, *Casting Deep Shade*, 31.
83. Wright and Moers, *Casting Deep Shade*, 31.
84. Wright and Moers, *Casting Deep Shade*, 31.
85. Wright and Moers, *Casting Deep Shade*, 22.
86. Wolfe, *What Is Posthumanism?* 119.
87. Wright and Moers, *Casting Deep Shade*, 64.
88. Wright and Moers, *Casting Deep Shade*, 65.
89. Wright and Moers, *Casting Deep Shade*, 24.
90. Wright and Moers, *Casting Deep Shade*, 25.
91. Wright and Moers, *Casting Deep Shade*, 71.
92. Wright and Moers, *Casting Deep Shade*, 71.
93. Wright and Moers, *Casting Deep Shade*, 71.
94. Wright and Moers, *Casting Deep Shade*, 79.
95. Wright and Moers, *Casting Deep Shade*, 72.
96. Wright and Moers, *Casting Deep Shade*, 72.
97. Wright and Moers, *Casting Deep Shade*, 73.
98. Wright and Moers, *Casting Deep Shade*, 74.
99. Wright and Moers, *Casting Deep Shade*, 74.
100. Wright and Moers, *Casting Deep Shade*, 239.
101. Haraway, *Staying with the Trouble*, 31.
102. Wright and Moers, *Casting Deep Shade*, 241.

103. Knickerbocker, *Ecopoetics*, 2.
104. Wright and Moers, *Casting Deep Shade*, 203.
105. Wright and Moers, *Casting Deep Shade*, 207, 210, 223, 241.
106. Wright and Moers, *Casting Deep Shade*, 67.
107. Wright and Moers, *Casting Deep Shade*, 50.
108. Wright and Moers, *Casting Deep Shade*, 40.
109. Wright and Moers, *Casting Deep Shade*, 196.
110. Wright and Moers, *Casting Deep Shade*, 235.
111. Ingold, "Bringing Things to Life," 1.
112. Wright and Moers, *Casting Deep Shade*, 2.
113. Wright and Moers, *Casting Deep Shade*, 211.
114. Ingold, "Bringing Things to Life," 10.
115. Wright and Moers, *Casting Deep Shade*, 243.

Acknowledgments

The idea for this article emerged from my work as a research assistant for Anna Poletti. I would like to express my sincere gratitude to Professor Poletti for her expertise and generosity throughout the supervisory and editing process.

Disclosure Statement

No potential conflict of interest was reported by the author(s).

Works Cited

Barad, Karen. *Meeting the Universe Halfway: Quantum Physics and the Entanglement of Matter and Meaning*. Durham, NC: Duke UP, 2007.

"Casting Deep Shade by C. D. Wright." *Publishers Weekly*, December 17, 2018. https://www.publishersweekly.com/978-1-55659-548-6

Derrida, Jacques. "Demeure: Fiction and Testimony." In *The Instant of My Death/Demeure: Fiction and Testimony*, edited by Elizabeth Rottenberg, 15–43. Stanford, CA: Stanford UP, 2000.

Felman, Shoshana, and Dori Laub. *Testimony: Crises of Witnessing in Literature, Psychoanalysis and History*. New York, NY: Routledge, 1991.

Fishman, Eric. "Poetry Review: 'Casting Deep Shade'—on Humanity and the Beech Tree." *Arts Fuse*, March 22, 2019. https://artsfuse.org/182866/poetry-review-casting-deep-shade-on-humanity-and-the-beech-tree/

Gilmore, Leigh. *Tainted Witness: Why We Doubt What Women Say about Their Lives*. New York, NY: Columbia UP, 2018.

Haraway, Donna. *Staying with the Trouble: Making Kin in the Cthulucene*. Durham, NC: Duke UP, 2016.

Ingold, Tim. "Bringing Things to Life: Creative Entanglements in a World of Materials." National Centre for Research Methods Working Paper no. 15, July 2010. Manchester: University of Manchester.

Knickerbocker, Scott. *Ecopoetics: The Language of Nature, the Nature of Language*. Amherst, MA: U of Massachusetts P, 2012.

Marshall, Lamar, and Rickey Walker. *Indian Trails of the Warrior Mountains.* Lawrence County, AL: Lawrence County School, 2005.

Muske-Dukes, Carol. "A Poet and Essayist Meditates on the Deep Roots Binding Humans and Trees." *New York Times*, April 5, 2019. https://www.nytimes.com/2019/04/05/books/review/cd-wright-casting-deep-shade.html

Tsing, Anna. *Mushroom at the End of the World: On the Possibility of Life in Capitalist Ruins.* Princeton, NJ: Princeton UP, 2017.

Whitlock, Gillian. "Post-ing Lives." *Biography* 35, no. 1 (2012): v–xvi.

Wolfe, Cary. *What Is Posthumanism?* Minneapolis, MN: U of Minnesota P, 2011.

Wright, C. D., and Denny Moers. *Casting Deep Shade: An Amble Inscribed to Beech Trees and Co.* Port Townsend, WA: Copper Canyon, 2019.

Xie, Jenny. "Restoration in the Attention Economy: On Reading C. D. Wright's *Shallcross.*" *American Poets* 23, no. 1 (2018): 1–6.

Writing the Lives of Fungi at the End of the World

By Alexis Harley

ABSTRACT
This essay examines how three recent monographs writing the lives of fungi use the co-constitutive entanglements of mycorrhizal fungi and their symbionts in order to bring the contingent, relational conditions of being itself into sharper relief. The Anthropocene demands a heightened awareness of multispecies entanglements, bringing into question the humanist ideal of human agency by turning to the co-constitutive relations of human and nonhuman lives.

The Anthropocene forces into popular consciousness an awareness of unexpected relationships: the flying of airplanes has become implicated in the starvation of polar bears, chemical agriculture in a hurricane off the coast of Barbados, the popularity of the paleo diet, and the devastation of frog populations. In this, the phenomena of the Anthropocene demonstrate on a global scale that beings are constituted by their relatings—a relational ontology that ecologists have more traditionally illustrated on the scale of the local ecosystem. This essay explores how contemporary "nature writing" (a branch of life writing, as Alfred Hornung has it) contributes to the narration of relational ontology as a condition of, and in relationship to, the Anthropocene.[1] Three recent monographs about fungi—Nicholas P. Money's *Mushrooms: A Natural and Cultural History* (2017), Alison Pouliot's *The Allure of Fungi* (2018), and Anna Lowenhaupt Tsing's *The Mushroom at the End of the World: On the Possibility of Life in Capitalist Ruins* (2015)—illuminate a larger emerging trend of writing nonhuman lives in terms, at least in part, of their co-constitutive relations with human cultures. Recent writers about fungi represent their mycelial subject as particularly well placed to illustrate co-constitutive relations. Peter McCoy, for instance, opens *Radical Mycology* with the claim that "[a]ll life is interconnected. This is the primary lesson that fungi teach … mushrooms and other fungi permeate the world, connecting and turning its innumerable cycles to demonstrate that every act carries an immeasurable chain of effects."[2] Pouliot focuses her study "less on individual species and more on relationships. Between fungi and fungi. Fungi and plants. Fungi and animals including *Homo sapiens*. Fungi and soil. Fungi and life."[3] Tsing's *The Mushroom at the End of the World* is ostensibly the story of a single life form—its titular mushroom, the matsutake—but it turns out to be the story of its relations too: it is the life story of the migratory buzzards that come from Siberia to Japan looking for the insects and frogs that have evolved to live with the trees that are sustained by peasant forestry practices, which also foster matsutakes; it is the story of the Laotian migrants foraging in the Oregon state forest for matsutakes to sell to the importers for the Japanese gift market; it is the pines of Finland; it is the dietary aesthetics of its author. The already overtly relational fungi that colonize these texts facilitate a fresh apprehension of the immeasurable entangledness that is the nature of nature and, more particularly, a fresh apprehension of the entanglements that characterize the Anthropocene. But they also problematize a common implication of the term "Anthropocene"—namely, that the relationship between humans and the rest of the world operates in one direction only, with the agential human acting on inert nature. As these texts show, fungal nature, in its ramifying intra-relationships with plants and animals, climate and soil, culture and history, is far from inert.

The three works of ethnomycology discussed here have been published in the thick of a veritable fungus craze, with fungi now regularly mooted as potential solutions to the problems of the Anthropocene. Sustainability forums are abuzz with talk of "mycoremediation," which promises the rehabilitation of soils poisoned by industrial waste; of edible mushrooms grown on capitalism's waste products (from spent coffee grounds to waste paper); of a reishi-mushroom extract being trialed as an antiviral treatment for beleaguered honeybees; and of the ways in which ectomycorrhizal fungi (species that form symbiotic partnerships with plants by latching around their roots and vastly extending their capacity to source mineral nutrients) contribute to carbon sequestration (a process that we have recently learnt is jeopardized by agricultural overuse of nitrogen-based fertilizers).[4] The subtitles of Paul Stamets' *Mycelium Running—How Mushrooms Can Help Save the World*—and Tsing's *The Mushroom at the End of the World—On the Possibility of Life in Capitalist Ruins*—indicate this trend and, indeed, *Mycelium Running* opens with the claim that "mycological membranes … heal habitats suffering from poor nutrition, stress, and toxic waste. In this sense, mushrooms emerge as environmental guardians in a time critical to our mutual evolutionary survival."[5] Stamets proposes a United Organization of Organisms or UOO ("pronounced 'uh-oh'"), in which the species that "irresponsibly exploit[s] the Earth" should and does face "the possibility of being rejected by the biosphere as a virulent organism."[6] It is some indication of how mainstream the turn to fungus-as-savior has become that Stamets' (and mushrooms') messianic status is honored in the *Star Trek: Discovery* series (2017–), in which the fictional figure of Lieutenant Commander Paul Stamets, an astromycologist, is awarded the Starfleet Medal of Honor for saving the day through fungal technology.

Despite the volume of all this buzz, it is, in the English-speaking West at least, almost entirely postcedent to Paul Crutzen and Eugene Stoermer's declaration in 2000 that the Anthropocene is upon us, and perhaps even more recent than that. In 2013, Alison Pouliot and John Ryan introduced a journal issue on fungi and philosophy with the observation that "despite their ubiquity and ecological importance, fungi are largely unregarded, especially in English-speaking cultures where mycophobia is the overwhelming norm."[7] It is a complaint common to recent commentaries on fungi. McCoy notes that mycology is a neglected science and that "those in the west never hear, learn, think, or talk about fungi," so that "these organisms become increasingly easy to ignore and buried beneath layers of maligned mystery."[8] In *The Allure of Fungi*, Pouliot likewise points out that this neglect has potentially dire implications for the conservation of fungus species: the Australian government's central piece of environmental-protection legislation, she notes, lists around two thousand species of plants and animals but only two species of fungi.[9] This history of (Western) inattention to fungi is perhaps one of the reasons why stories of fungi as victims of the Anthropocene (and possible collaborators in surviving it) have an extra piquancy: the thing much ignored that turns out to matter a great deal is perhaps *the* central trope of Anthropocene narratives, which repeatedly draw out the ways in which causal networks defy human intentions or foreknowledge.

The historical neglect of fungi, the failure to recognize the complexity of co-constitutive entanglements, and a blindness to the limits of human control share knotted, unravelable origins with humanist autobiography—and also with the Anthropocene. The "onset of carboniferous industrialization" occurred in late-eighteenth-century Europe, and Crutzen has proposed a possible inaugural date for the Anthropocene with the patenting of the steam engine in 1784.[10] The rise of carboniferous industrialization is connected, via middle-class economic enfranchisement, with the conditions that also saw the ascendancy of humanism, with its resharpened awareness of the human|nature, mind|matter, subject|object, self|other divide. I am wary of providing a reductive causal account of how these phenomena—industrialization and humanist individualism—relate, particularly as I argue that the narratives of the Anthropocene chasten us for our limited insight into complex causative knots, but I am happy to let Jayne Lewis ask for me: "What, if not the Enlightenment, can we thank—or revile—for the technologies and habits of self-absorption to which we attribute climate change?"[11] We can certainly find evidence that nonhumans fared badly, rhetorically at least, under Enlightenment humanism. Human rights proponents of the 1790s avoided the inevitable tautology of "humans have rights because they are human" by arguing that humans have rights because they are not *not*-human—Mary Wollstonecraft's indignant differentiation between rights-bearing women and "gentle, domestic brutes" in *A Vindication of the Rights of Woman* and Thomas Paine's call for nations to "be rational, and not be governed like animals" in *Rights of Man* are two of many cases in point.[12] The emerging discourse of the rights of man and the humanist individualism that would make the story of a common person a subject of public interest collide in the rise of autobiography. It is no coincidence that one of the most famous autobiographers in European history, Jean-Jacques Rousseau, is also the author of a foundational document in the philosophy of human rights, *Du contract social* (1762).[13]

Autobiography—a word that first appeared in German in the mid-eighteenth century and in English in the 1790s[14]—became in the eighteenth and nineteenth centuries, and beyond, one of the most significant sites for the inscription of this free-willing self, reified by its own consciousness. As Felicity Nussbaum puts it, humanist ideology supposes that "each individual participates in a transhistorical human nature that is distinguished from the nature of other species by its ability to make ethical decisions and its reverence for the past."[15] To call this individual's participation in "human nature" "transhistorical" is to say that the nature of being human is to be radically *un*entangled with the conditions (be they cultural, climatic, or organic) in which one exists. The humanist's human floats freely above the particularities of time and space and matter. It is thus in the thick of the late European Enlightenment that life writing starts to cleave apart into, on the one hand, genres that represent human life and, on the other, those that deal with other-than-human life. The words "biology" and "biography" are close cousins and were originally far more interchangeable than they are today. In the late-seventeenth century, "biology" related to what we would now call biographical study and writing. Although this use was still occasionally in play as late as 1813, when James Field Stanfield paired biology (biographical research) with biography (biographical writing),[16] by the end of the eighteenth century, these words had clearly begun to speciate. Biography was to become the narrative of an individual (human) life and biology a branch of science dealing with living (mostly nonhuman) organisms—one to do with individuals, *persons*, their stories; the other to do (mostly) with life in general rather than the individual life. But even biological life writing of the eighteenth century was infused with the spirit of individualism, a mode that retarded apprehension of the relational ontology of organic life, and undermined understanding of fungi in particular.

In *The Allure of Fungi*, Pouliot records with some glee the blackening squishy mess made by dozens of ink-cap mushrooms on the otherwise neat lawns of the Linnaeus Garden in Uppsala, Sweden.[17] From the 1730s until his death in 1778, Swedish natural historian Carl von Linné refined his taxonomy of the organic world, classifying almost eight thousand plants and more than four thousand animal species—but only eighty species of fungi.[18] Modern biologists have assigned an entire kingdom to fungi (of which there may be more than five million species) but, for Linnaeus, fungi were a motley minor assortment of not-quite plants to be filed under "algae," itself one of the suborders of *cryptogamia* (Greek for "concealed marriages"), which included all "plants" without flowers.

Fungi were an accidental casualty of Linnaeus's cultural bias: he happened to be a mycophobe, who claimed in his *Flora Lapponica* (1737) that only foreigners ate mushrooms.[19] But it also turns out that an understanding of fungal lifeways was particularly ill-served by Linnaeus's configuration of species in discrete taxa that foreground the relatedness of similar species but conceal the relatedness of dissimilar ones. As we now know, a substantial proportion of fungal species live in symbiotic or parasitic relationships with other very different species; their existence hinges on (and sometimes enables) the existence of their plant or animal symbionts or hosts. Linnaeus led a revolution in the representation of life, regimenting the messy diversity of life forms—emphasizing the closeness of similar-looking forms (particularly emphasizing similarities between reproductive mechanisms) while occluding the relationship between very different ones. While Linnaeus's system undoubtedly helped later evolutionary theorists, like fellow botanist Erasmus Darwin, to find the evidence for similar species' shared ancestry, it also obscured the lateral co-constitutive relationships between species: had Linnaeus classified the mycorrhizal pine mushroom *Lactarius deliciosus*, which depends for sugars on the pine tree *Pinus radiata*, he would have placed it as far away as possible from the pine tree, which, in turn, depends on its fungal symbiont to melt through hard rock and deliver essential minerals—and, just as far away, the *Homo sapiens*, who clear-fells vast tracts of forest to establish pine plantations, and harvests both mushrooms and timber.

Pouliot's joy at the triumph of the Swedish ink caps over the neatly clipped lawn probably owes more to her feelings about Linnaeus's renowned personal mycophobia than to her feelings about his classificatory system. Indeed, as she writes, "[n]aming … gives a fungus a place in the great scheme and hierarchy of taxonomic mycology. Having a name increases the likelihood of inclusion—in human consciousness, in research and in conservation programs."[20] In other words, in a human-dominated world where species' futures depend on human recognition that they matter, co-option into a Linnaean schema is a rhetorical advantage. On the other hand, she also argues that organisms that seem to "comprise discrete individual entities" are never "truly individual," "given the staggering array of organisms performing vital processes in their body cavities."[21] Cue Michael Pollan's memorable remark about having his microbiome sequenced by the American Gut Project: "I can tell you the exact date that I began to think of myself in the first person plural—as a superorganism, that is, rather than a plain old individual human being."[22] On another front, Pouliot notes that there is an artificiality to imposing a "species identity" on plural lives, particularly lives like those of fungi that make trouble for such assimilating categories as species. Is it really possible to lump into one category the twenty-eight thousand (and counting) sexually compatible genders (or "mating types") of the *Schizophyllum commune* "species"?[23] In making these two points, Pouliot worries at both ends of the taxonomizing process: a taxon is inevitably too broad (flattening a vast array of differences in order to create the illusion of species identity) and too narrow (repressing the intrinsic intra-relationship, co-making, or dependence of "species" with each other). Ultimately, like all the writers of fungal life discussed in this essay, Pouliot is concerned to document the relationship between (human) culture and fungi, and thus she cycles through the effects of conventional (and new molecular) classifications of fungi on how they are represented and treated, as well as how fungi are positioned by the attitudes of semi-professional foragers, farmers interested in how fungi can increase their crop yields, and the fungal epistemology of the Wiradjuri people (traditional custodians of lands in what is now called central New South Wales, Australia), whose names for fungi draw out their kinship with other inhabitants of places and country itself. Throughout her book, Pouliot also explores her own relationship to fungi—a relationship that she allows has changed how she walks in the forest, how she observes, and how she thinks about the nature of nature. The autobiographical gestures in Pouliot's writing decenter the human self. She represents herself as much as the object of fungi's influence as the influencing agent, a turn likewise practiced by Tsing: "The smell of matsutake," she writes, "transformed me in a physical way."[24]

Despite my suggestion above that biography and biology parted ways in the eighteenth century, one representing the individual and the other life in general, the ascendancy of Linnaeus's representation of life in discrete taxa bent biological thinking toward a kind of individualism too. The difficulty in pinning down fungal organisms as individuals—in the same way that humanist life writing attempts to pin down human individuals for representation—is perhaps one of the reasons for the long-standing neglect of fungi. Why is there this difficulty in configuring fungi as objects? Firstly, as Money shows in *Mushrooms*, the fungal individual is not what many of us think it is. The mushroom itself "is not a self-contained organism, like a jellyfish, for example"; rather, "it is a reproductive organ produced by a colony, or mycelium, that grows in soil or rotting wood."[25] In a memorable comparison, Money notes that to confuse the mushroom with the mushroom-forming organism "is a bit like using a photograph of a large pair of testicles to represent an elephant."[26] Like many attempts to compare fungi and animals, this one flounders, in that an elephant's testicles are part of its bodily identity from its gestation until its death, barring accidents; the mushroom-forming organism, on the other hand, does not grow its "testicles" until, as a fully mature organism, it has fused subterranean hyphae with a compatible mating organism. Conversely, the cell in an elephant's testicle does not contain all the genetic material needed to grow a new elephant; the spore in a mushroom, on the other hand, is ready to begin a new organism, so long as it is catapulted onto a propitious landing spot. It is worth enlarging on the limits of the mushroom|testicle analogy because it demonstrates something of the special indeterminacy of fungi—the difficulty in making them conform to human, or at least animal, norms, which is another reason why they challenge representation as individuals. Tsing contrasts the "indeterminacy of fungal growth" with human bodies that "achieve a determinate form early in our lives." In the normal course of events, "we can't grow extra limbs, and we're stuck with the one brain we've each got. In contrast, fungi keep growing and changing form all their lives. Fungi are famous for changing shape in relation to their encounters and environments."[27] In this comparison, fungi perhaps seem secondary to humans: human forms are determinate, but fungal forms are *other*—indeterminate. However, not to fall into a kind of binary thinking that makes human experience normative, Tsing asks how humans might, in fact, be compared with fungi: "What if our indeterminate life form was not the shape of our bodies but rather the shape of our motions over time? Such indeterminacy expands our concept of human life, showing us how we are transformed by encounter. Humans and fungi share such here-and-now transformations through encounter."[28]

This relational ontology is the condition of all life, perhaps all matter, but it is particularly overt in the case of symbionts, organisms of different species that depend on one another in the most immediate of ways. One of the reasons why fungi work so effectively to illustrate relational ontology is the reputation of mycorrhizal fungi as some of the most remarkable symbionts on earth (though it is worth noting that some mycorrhizal fungi are better thought of as parasites than symbionts). There are, Money estimates, around eight thousand known species of mycorrhizal associations between fungi and plants. A symbiotic mycorrhizal association occurs when the mycelium sheathes a plant's roots and vastly extends the root system's reach. The mycelium provides the plant with hard-to-access minerals and a layer of protection against would-be parasites, and may enable communication and nutrient exchange with other plants, even connecting whole forests in what has been referred to by almost everyone who writes on the subject as a "woodwide web." In exchange, the plant provides the mycelium with sugars (indeed, possibly up to 80% of a plant's sugars are produced to support its symbionts). These relationships are essential to the existence of both trees and fungi, and so also to those that live in forests and—another Anthropocene lesson—anything that benefits from the carbon-sequestering work that forests do.

Mycorrhizal associations produce, as Tsing puts it, "an infrastructure of interspecies communication."[29] It is an enchanting parable of mutualism, and yet "the world-building work of fungi" has been for a long time ignored. Tsing attributes this to the fact that "until quite recently many people—perhaps especially scientists—imagined life as a matter of species-by-species reproduction." Crucially, "[m]utualistic relations were interesting anomalies, but not really necessary to understand life." Rather, "[l]ife emerged from the self-replication of each species, which faced evolutionary and environmental challenges on its own. No species needed another for its continuing vitality; it organised itself"—a view that Tsing calls a "self-creation marching band," which overwhelms "the stories of the underground city."[30] Tsing cites the work of Scott Gilbert and his colleagues on the so-called "hologenome theory" of evolution. This work proposes that "[a]lmost all development may be codevelopment. By codevelopment we refer to the ability of the cells of one species to assist the normal construction of the body of another species." Instead of thinking of life in terms of "internally self-organizing systems, self-organized through 'autopoiesis,'" this work proposes that "symbiosis appears to be the 'rule', not the exception. … Nature may be selecting 'relationships' rather than individuals or genomes.'"[31]

The central character of *The Mushroom at the End of the World* is the matsutake mushroom or mushroom-forming organism. Tsing follows the matsutake around the world, showing how matsutakes respond tactically to environmental pressures, particularly the pressures that relate to two of the parties with which matsutakes are most intimately entangled: humans and trees. Matsutakes form mycorrhizal associations principally with pine trees. Tsing argues that the matsutake fungus tracks "weather changes from year to year" and engages in "multiyear strategic planning so that carbohydrates stored up one year might be expended in later fruiting." She also suggests that trees with mycorrhizal partners do the same thing. This "masting," as periodic fruiting is called, appears "to be coordinated between trees and their fungi"—fungi storing carbohydrates for the fruiting of trees and trees accommodating "the uneven fruiting of fungi."[32] Crucially, Tsing is implying that the question of whether to mushroom or not is the subject of fungal (or fungal-plant) deliberation, in a weighing up of resources, threats, and social obligations. This mycorrhizal intelligence that takes place in the contingency of interspecies relationships suggests that intelligence—intentionality, agency, mind—need not be housed in the nucleus of a self-actualizing individual. It can exist between. And it raises questions about what counts as human intelligence, intentionality, and agency.

The surprise twist in the Anthropocene history of the matsutake is that the mushroom's fortunes have suffered not because of human disturbance, but because of the lack of it. The mushrooms require pine trees. In Japan, as Tsing explains, pine trees will only grow if gaps are made in the forest, if indigenous broad-leaved trees are removed, so that pines can get the sunlight they need. When Japanese peasants stopped chopping down trees for fuel, no one intended, or imagined, the demise of the matsutake. It turns out that people are entangled with mushrooms as more than just consumers—indeed, people are as much invested as the pine trees themselves in the nutrient exchange with matsutakes. The anthropological insights of *The Mushroom at the End of the World* seem to be informed by the contingent, potentially mutualistic relations between fungi, humans, and trees, applying the principle of these relations to interactions of economic and social planes. Examining the mushroom-foraging camps of Oregon, where it is mostly Asian migrants who collect American mushrooms for export to Japan, Tsing describes in detail the social dynamics of mushroom-pickers negotiating buying conditions. The web of relationships and nutrient exchange expands to encompass less tangibly physical forms of exchange: "what is exchanged every evening is not just mushrooms and money. Pickers, buyers, and field agents are engaged in dramatic enactments of freedom, as they separately understand it, and they exchange these, encouraging each other, along with their trophies: money and mushrooms. Sometimes, indeed, it seemed to me that the really important exchange was the freedom, with the mushroom-and-money trophies as extensions—proofs, as it were—of the performance."[33] The "nutrients" here are as much psychological (the impression of freedom, though a freedom that exists through relationship, through contingent negotiation with another) as they are material (the mushrooms that are food) and symbolic of matter (the money is a symbol for the material goods it could potentially buy). The several overlapping planes of entanglement here play out in another scene of complex entanglement between ecological, economic, material, psychological, and social sites of interrelationship: "The surprising thing about the search for matsutake in Oregon is that it does not involve alienation in the relation between foragers and mushrooms. The mushrooms are indeed torn from their fungal bodies (although, as fruit, this is their goal). But instead of becoming alienated commodities, ready for conversions between money and capital, they become trophies of the hunt—even as they are sold. Foragers beam with pride while showing off their mushrooms; they can't stop narrating the pleasures and the dangers of the search. The mushrooms become part of the foragers, just as if they had eaten them."[34] The mushrooms are also, or soon to become, capitalist commodities. But that term seems to imply that they are the passive objects of others' transactions, rather than that these transformative relationships are an effect of the fungal bodies' "goal"—to have their spore dispersed far and wide by foragers.

The idea that fungi might have "goals" sits uneasily alongside the prevailing sense of mushrooms as vegetally inert, but fungi repeatedly demonstrate their capacity for planning and problem-solving. In a chapter on the history of mushroom cultivation, Money relays from Song dynasty records the methods used to induce the mushrooming of shiitakes (*Lentinula edodes*). Shiitakes digest dead hardwood trees and logs (their Japanese name derives from one of their favorite foods, the shii tree, *Castanopsis cuspidata*), converting highly complex carbohydrates into forms more available to other microorganisms. Left to their own devices, the mycelium can go for years without "fruiting" or producing mushrooms. But the shiitake cultivators of the Song dynasty urged on mushroom production by burying logs for a year, then unearthing them, dunking them in cold water, and slamming them with wooden clubs. It seems that the temperature shock and the thumping trigger mushroom emergence as the mycelia rally the resources they need to move their genes on to a nicer neighborhood. By 2011, Money relates, China had come to control eighty percent of a US$20 billion mushroom market, and Japan's domestic production of shiitake was in a fifty-thousand-kilogram-per-annum deficit; unsurprisingly, since then, researchers have been investigating more effective mycelium stressors than the log-contusion technique. Recent Japanese experiments into the effects of simulated lightning strikes on eight species of saprophytic fungi, including *Lentinula edodes*, show that when logs colonized with shiitake mycelia receive an electric shock, they withdraw energy from their submerged hyphae and instead invest in the production of mushrooms—in fact, more than twice the number of mushrooms ventured up by the control mycelia.

Unless they are eaten first, Money tells us, these mushrooms are extraordinarily effective propaganda machines, their gills flinging spores at a rate of more than 100,000 times the spores' own length per second. The spores are genetically different from the two compatible mating organisms that make the mushroom (in some cases, there is even genetic difference between spores within the same mushroom). These proliferating genetic differences, of course, give the family more chances of surviving in the changed environments into which the youngsters are dispersed; just as importantly, the catapulting of spores into the air enables the fungus to send its offspring off to more possible refuges, away from the current one that is being zapped with a cattle prod.

The research strongly implies that shiitakes are fruiting more and more, in a kind of generative anxiety around the existential threat to their substrates. In contingent relationship with those that share their environment, these fungal bodies are demonstrating their desire to survive and their capacity to strategize.

Away from the somewhat contrived stresses of the shiitake farm, Money relays the findings of a pair of research projects that show how various species of mushroom-producing fungi have both responded and potentially contributed to climate change. In southwest England, warmer summer temperatures and more autumn rainfall have extended the autumn mushroom season so that it has more than doubled in length since the 1950s. A number of wood-decay (rather than mycorrhizal) fungal species in this region, formerly "fruiting" only once a year, are now sending up mushrooms in both autumn and spring. This indicates a prolongation in the period of annual growth of the mycelium (the submerged filamentary organism that makes mushrooms) and impacts on soil fertility, plant development, and the availability of food for wild animals.[35] It also means increased carbon dioxide emissions. Like animals, fungi respire carbon dioxide and the more wood-based carbon they consume, the more carbon dioxide they emit. These fungi are affected by the weather of increased atmospheric carbon, but it seems that they also effect it by munching more rapidly through the available wood supply and releasing more atmospheric carbon themselves. In another astonishing story of the complex interactions of fungi and weather, Money explains the role of mushroom spores in cloud-seeding above rainforests. If there is not enough humidity to prompt the mushrooms to spore, the clouds will be less likely to form, and so it goes on,[36] in a potentially catastrophic iteration of what is often referred to as a "feedback loop."

This term reduces the complex interactions between agents to a kind of blind algorithm, refusing the "effect" the status of a "cause" (instead it is represented as an effect *en route* to further effects). But Money's examples indicate that the ways "nature" has responded to climate change blur the delineation between cause and effect, agent and patient, subject and object. "The point of living in the epoch of the anthropocene," Bruno Latour suggests, "is that all agents share the same shape-changing destiny."[37] The closer we look into the stories of the Anthropocene, the more we are required to revise how we imagine the nature of change, to apprehend the interknotted play of agencies, human and more-than-human, and to understand how agency can operate outside the model of the rational controlling intelligence whose impacts proceed predictably in line with its intentions.

The knotty interactions of agents implicated in "the same shape-changing destiny" are elucidated in these mushroom books. In an Anthropocenic world, there can, of course, be no present-tense natural history that depicts nonhuman lives which are uninfluenced by humans, because even in the most untrampled wilderness, the effects of human behavior are present in the very weather. As Emma Marris puts it, "[n]ature is almost everywhere. But wherever it is, there is one thing that nature is not: pristine. … We've been changing the landscapes we inhabit for millennia, and these days our reach is truly global. Inhale. That breath has 36 percent more molecules of carbon dioxide than it would have had in 1750."[38] The global penetration of human influence is what "Anthropocene" means, after all. But I argue that these new histories of mushrooms yoke humans and fungi in ways that do not just accede to the ubiquity of human influence on nonhuman lives, but demonstrate how implicated fungi and humans are in each other's destinies.

The Anthropocene both affirms and disturbs Enlightenment humanism's idea of human agency. The term implies, as its founders have explained, that collective human activity is sufficient to drive the earth into a new geological epoch. While the grammar of this formulation might seem to confirm humanity's singular power over nature, and thus humanist ideals of human agency, closer inspection of what "Anthropocene" means for that agency suggests a radical threat to the humanist ideal. Rather than showing the human subject rationally commanding the object that is nature, the singular cause anticipating and controlling its singular, predictable effect, the Anthropocene has made manifest the ways the effects of human actions both exceed and fall short of human intentions. "Humanity" is an agent of the Anthropocene, yes, but this is not the intelligent, individually ordained agency that humanism's humans are supposed to exert. Latour has asked how humanity could "simultaneously be part of such a long history [of engendering climate change], have such an important influence, and yet be so late in realizing what has happened and so utterly impotent in our attempts to fix it."[39] The question, he suggests, demands a revised understanding of the nature of our agency: humans' "active role" in the existence of the phenomena of the climate crisis does not bespeak agency with its commonly understood valency of intentionality or "authorial control." Our "agency" has for the large part been, instead, unwitting and inadvertent; humans, in Latour's formulation, are "subjects subjected to the vagaries of their own interconnected collisions," the Anthropos "playing the role of the dumb object."[40] Meanwhile, Latour argues, the sometimes "dumb object" on which humans are supposed to act—the earth, nature, matter—is itself revealed as a new kind of subject, a "subject" because it is *subjected* to another's action,[41] in which subjection is itself active, changing, and responsive in ways that the people to whose actions it is responding never intended and failed to anticipate.

Biography and biology have been separated, I noted above, but the spirit that moved their separation has not done humans the good it was supposed to. Timothy Morton claims that the Anthropocene means that "we are no longer able to think history as exclusively human" for "in this period, nonhumans make decisive contact with humans, even the ones busy shoring up differences between humans and the rest."[42] The recoupling of natural and cultural history afoot in the books studied here addresses the reality that the bulwark "between humans and the rest" is porous and futile, that our history is not exclusively human (but utterly implicated with the lives all around and in and with us), and likewise that the history of "nature" is not exclusively "natural." In the Anthropocene, but really always in any case, natural history is also cultural history.

LaTrobe University

Notes

1. Hornung, *Ecology and Life Writing*, x.
2. McCoy, *Radical Mycology*, xv.
3. Pouliot, *The Allure of Fungi*, 2.
4. Averill, Dietze, and Bhatnagar, "Continental-Scale Nitrogen Pollution."
5. Stamets, *Mycelium Running*, 1.
6. Stamets, *Mycelium Running*, 1.
7. Pouliot and Ryan, "Fungi," 1.
8. McCoy, *Radical Mycology*, xvi.
9. Pouliot, *The Allure of Fungi*, 242.

10. Ford, "Punctuating History circa 1800," 78; Crutzen, "The 'Anthropocene,'" 13.
11. Lewis, "Change Beyond Belief," 112.
12. Wollstonecraft, Vindication, 33; Paine, *Rights of Man*, 118.
13. Rousseau, *Du contrat social; ou Principes du droit politique.*
14. "Autobiography."
15. Nussbaum, "The Politics of Subjectivity," 160.
16. Stanfield, *Essay*, xi.
17. Pouliot, *The Allure of Fungi*, 155.
18. Pouliot, *The Allure of Fungi*, 155.
19. Bertelsten, *Mushroom*, 23.
20. Pouliot, *The Allure of Fungi*, 156.
21. Pouliot, *The Allure of Fungi*, 154.
22. Pollan, "My Best Friends."
23. Pouliot, *The Allure of Fungi*, 155.
24. Tsing, *Mushroom*, 47.
25. Money, *Mushrooms*, 7.
26. Money, *Mushrooms*, 8.
27. Tsing, *Mushroom*, 47.
28. Tsing, *Mushroom*, 47.
29. Tsing, *Mushroom*, 139.
30. Tsing, *Mushroom*, 139.
31. Gilbert et al., "Symbiosis," 673.
32. Tsing, *Mushroom*, 174–175.
33. Tsing, *Mushroom*, 75.
34. Tsing, *Mushroom*, 121.
35. Money, *Mushrooms*, 169.
36. Money, *Mushrooms*, 170.
37. Latour, "Agency," 17.
38. Marris, *Rambunctious Garden*, 2.
39. Latour, "Agency," 2.
40. Latour, "Agency," 6, 13.
41. Latour, "Agency," 5.
42. Morton, *Hyperobjects*, 5.

Disclosure Statement

No potential conflict of interest was reported by the author(s).

Works Cited

"Autobiography." *Oxford English Dictionary*. Accessed November 3, 2019. https://www.oed.com/

Averill, Collin, Michael C. Dietze, and Jennifer M. Bhatnagar. "Continental-Scale Nitrogen Pollution Is Shifting Forest Mycorrhizal Associations and Soil Carbon Stocks." *Global Change Biology* 24, no. 10 (2018): 4544–4553.

Bertelsten, Cynthia. *Mushroom: A Global History*. London: Reaktion, 2013.

Crutzen, Paul J. "The 'Anthropocene.'" In *Earth System Science in the Anthropocene*, edited by Eckart Ehlers and Thomas Krafft, 13–18. Berlin: Springer, 2006.

Ford, Thomas H. "Punctuating History circa 1800: The Air of *Jane Eyre*." In *Anthropocene Reading: Literary History in Geologic Times*, edited by Tobias Menely and Jesse Oak Taylor, 78–95. University Park, PA: U of Pennsylvania P, 2017.

Gilbert, Scott F., Emily McDonald, Nicole Boyle, Nicholas Buttino, Lin Gyi, Mark Mai, Neelakantan Prakash, and James Robinson. "Symbiosis as a Source of Selectable Epigenetic Variation: Taking the Heat for the Big Guy." *Philosophical Transactions of the Royal Society B* 365, no. 1540 (2010): 671–678.

Hornung, Alfred. *Ecology and Life Writing*. Heidelberg: American Studies, 2013.

Latour, Bruno. "Agency at the Time of the Anthropocene." *New Literary History* 45, no. 1 (2014): 1–18.

Lewis, Jayne. "Change Beyond Belief: Fictions of (the) Enlightenment and Simpson's 'Climate Change Suite.'" In *A Cultural History of Climate Change*, edited by Tom Bristow and Thomas H. Ford, 111–127. London: Routledge, 2016.

Marris, Emma. *Rambunctious Garden: Saving Nature in a Post-Wild World*. New York, NY: Bloomsbury, 2011.

McCoy, Peter. *Radical Mycology: A Treatise on Seeing and Working with Fungi*. Portland, OR: Chthaeus, 2016.

Money, Nicholas P. *Mushrooms: A Natural and Cultural History*. London: Reaktion, 2017.

Morton, Timothy. *Hyperobjects: Philosophy and Ecology after the End of the World*. Minneapolis, MN: U of Minnesota P, 2013.

Nussbaum, Felicity. "The Politics of Subjectivity." In *Women, Autobiography, Theory: A Reader*, edited by Sidonie Smith and Julia Watson, 160–167. Madison, WI: U of Wisconsin P, 1998.

Paine, Thomas. *Rights of Man, Part the Second: Combining Principle and Practice*. London, 1792.

Pollan, Michael. "Some of My Best Friends Are Germs." *New York Times Magazine*, May 19, 2013. https://www.nytimes.com/2013/05/19/magazine/say-hello-to-the-100-trillion-bacteria-that-make-up-your-microbiome.html

Pouliot, Alison. *The Allure of Fungi*. Clayton, VIC: Commonwealth Scientific and Industrial Research Organisation, 2018.

Pouliot, Alison, and John Ryan. "Fungi: An Entangled Exploration." *PAN: Philosophy Activism Nature* 10 (2013): 1–5.

Rousseau, Jacques. *Du contrat social; ou Principes du droit politique*. Amsterdam: Marc Michel Rey, 1762.

Stamets, Paul. *Mycelium Running: How Mushrooms Can Help Save the World*. Berkeley, CA: Ten Speed, 2011.

Stanfield, James Field. *An Essay on the Study and Composition of Biography*. London: Sunderland, 1813.

Tsing, Anna Lowenhaupt. *The Mushroom at the End of the World: On the Possibility of Life in Capitalist Ruins*. Princeton, NJ: Princeton UP, 2015.

Wollstonecraft, Mary. *A Vindication of the Rights of Woman: With Strictures on Political and Moral Subjects*. London, 1792.

Planetary Delta: Anthropocene Lives in the Blues Memoir

By Parker Krieg

ABSTRACT

This article argues that blues memoirs are an example of life writing in the Anthropocene. Building on ecocritical scholarship which suggests that blues is a neglected source of environmental culture that reframes debates around race, economy, and culture, it asks how blues memoirs offer alternative perspectives on the Anthropocene.

> Common, everyday life, the background of our national life, is to be seen through the blues: trains, ships, trade unions, planes, the Army, the Navy, the White House, plantations, elections, poll tax, the boll weevil, landlords, epidemics, bosses, Jim Crow, lynchings.
>
> — Richard Wright, qtd. in Michel Fabre, *The Unfinished Quest of Ralph Ellison.*

Introduction

In *Darker Than Blue*, Paul Gilroy quotes Jean Toomer, who tells a journalist in 1931: "I am of no particular race, I am of the human race, a man at large in the human world, preparing a new race. I am of no specific region. I am of earth."[1] Toomer's claim on the universal, radical then as it is today, contains an all-too-familiar *geographical* formulation. That the past has a *place* while the future is *space*, nowhere in particular, or perhaps distributed everywhere on a smooth sphere without regional distinctiveness, speaks to the difficulties of the Anthropocene, an era whose belated recognition as a new global scale of periodization threatens to eclipse the uneven material geographies of those living it. These changed conditions do not resolve Toomer's dilemma, but rather pose the question anew: Is it possible to write *life* in the Anthropocene without reference to particular earthly regions and entanglements? Moreover, if there is an *anthropos* to come, where might we look for Toomer's "man at large in the human world" preparing its arrival?

This essay finds one possible answer to this question in the blues memoirs of artists from the Mississippi Delta. It traces environmentally inflected life writing across three texts: B. B. King's *Blues All Around Me*, W. C. Handy's *Father of the Blues*, and Willie Dixon's *I Am the Blues*. In asking what kind of Anthropocene lives are being written, I hope to challenge Anthropocene periodizations that exclude colonial and plantation geographies from this history, and to argue that blues texts can be read as a record of life writing in the Anthropocene which speaks to contemporary environmental culture. Blues expresses what Richard Mizelle, Jr. calls the "double environmental burden" of ecological precarity and racial inequality, complicating Romantic lineages of American environmental thought.[2] As autobiographical texts, they challenge what Paul John Eakin calls "the myth of autonomy" associated with the self-determined, transparent author of traditional autobiography.[3] This "relational self" is drawn from a blues epistemology that situates the blues memoir within a longer history of human-environment relations.[4] Emerging from the songs and gestures of colonized and enslaved workers, the pragmatic performativity of the "blues voice" lends itself to contemporary practices of reading autobiography that emphasize the performative, material, and discursive self.[5] In passages that highlight nonhuman agency in composition, material location within networks of production, and emotional ambiguity at the crossroads of the psychological, the social, and the natural, the blues dimensions of these memoirs present alternative forms of selfhood in the Anthropocene.

Anthropocene Lives, Writing

In 1970, Albert Murray predicted that "someday students of machine-age culture in the United States may find that Negro slaves in the cotton fields had already begun confronting and evolving aesthetic solutions for the problems of assembly line regimentation, depersonalization, and collectivization."[6] His remark turns the tables on the post-sixties image of blues as the rural expression of preindustrial life tailored to the romantic ideals of the white counterculture. Murray's link between the industrial logic of the plantation economy and that of Fordism identifies a historical continuity in forms of New World production. The colonial combination of field and factory, and its various returns in contemporary global capitalism, has led scholars to pose the "Plantationocene" as a framework for critically situating the economic, racial, and settler-colonial foundations of Anthropocene histories.[7] Bruno Latour argues for Plantationocene as a more productive title for the epoch's "historical 'de-soilization' of the Earth."[8] As it strives toward universal description, the Anthropocene concept symptomatically carries with it some of the same distinctions (for example, nature|culture) that have produced a world of disposable externalities. "It is striking," Latour continues, "how much analytical work is now needed to re-localise, to re-territorialise and re-earth, to 'reground,' basically, practice."[9] Geographer Clyde Woods traces similar relationships between environment, economy, and culture in his classic work *Development Arrested: The Blues and Plantation Power in the Mississippi Delta*. For Woods, the blues is an epistemology, a "longstanding African American tradition of explaining reality and change," which emerged in response to contradictory conditions of the plantation system and its post-Reconstruction continuation.[10] In this sense, the blues (and its tributaries) are an intimate expression of the Plantationocene, which is part of the larger unfolding critical and cultural narrative of the Anthropocene.

To consider the Anthropocene and life writing together is to raise questions of genre alongside geology. For some, "Anthropocene is a science-fiction concept, that is, a concept that pulls us out of familiar space and time to view our predicaments as if they belonged to a distant land."[11] However, Ralph Ellison's description of the "blues voice" as a performance that "mocks the despair stated explicitly in the lyric" reveals an older genre: "the great human joke directed against the universe, that joke which is the secret of all folklore and myth: that though we be dismembered daily we shall always rise up again."[12] These two genres—science fiction and blues—offer differing conceptions of the self in relation to space and time, as well as differing sets of cultural attitudes: the former offers the speculative high of *defamiliarization* or dislocation from a position of comfort, while the latter is a *diasporic* figure that recomposes itself after dispersion or injury. Put another way, Jason W. Moore argues that the metaphor of "spaceship earth" carries "slaveship earth" along with it.[13] These two figures of the planet map different social relations, indicating contradictory experiences *within* the Anthropocene. For philosopher Clive Hamilton, the subject that emerges in the Anthropocene is a *tragic* figure which recognizes its species-being as a planetary force as it simultaneously becomes aware of all the ways it is deeply vulnerable.[14] Yet this *newfound* precarity is unevenly distributed. As Laura Pulido observes, life in the Anthropocene is split along an axis of "differential vulnerability" as racialized subjects face both disproportionate geographic and material risks and expropriated benefits.[15] Blues is thus not a *mere* difference in genre, but a historical product of the very socioecological arrangements—the Plantationocene, the Capitalocene—that constitute the epoch. Blues operates on this axis of differential vulnerability as an expression of life writing within the long history of the anthropogenic present.

An example of this differential vulnerability is the Great Mississippi Flood of 1927, which affected a million people and displaced over half a million African American sharecroppers. As Susan Scott Parrish observes, "the contours of the 1927 disaster—its racial and regional manifestations, its political and media handling, its repercussions to human conceptions of self—show us what life within the Anthropocene has become."[16] The federal response reinforced the racial order of the plantation economy by imposing a "high modernist" approach to relief infrastructure and conscripted black labor (often at gunpoint) on the levees.[17] That spring, Bessie Smith's "Backwater Blues" became a chart-topping hit and national "eclogue" testifying to the sublime destruction of the 1927 floods, while speaking to a "multiracial [and] multiregional public" in a manner that "invited identification with a female refugee."[18] Smith's song narrates collective experience through an "I" that translates a national emergency into a blues idiom. Yet the song's actual history breaks the assumed "autobiographical pact" between the singer, lyrics, and flood event, and instead suggests a performative mode of life writing in an uneven landscape.[19] In *Backwater Blues: The Mississippi Flood of 1927 in the African American Imagination*, Mizelle describes how Smith wrote the song months earlier when her Southern tour was disrupted by segregated transportation in flooded "backwater" tributaries and towns.[20] Its release two months before the main flood event might be read as an example of the differential vulnerability manifested in geographic and racial (im)mobility, as conditions were already being felt *before* the national event. What enabled Smith's "prescience" was the fact that her narrated "I" had confronted a racialized environmental disaster long before the song became a national hit, eliciting sympathetic identification from multiethnic audiences throughout the country.

The autobiographical self of flood songs is joined by personal accounts of journalists who went down to cover the governmental response, concerned that new developments marked the return of old horrors. Wishing to control the movement of their indebted sharecroppers, the planter class allied with government forces to hold many of them as conscripted labor on the levees. The confluence resembles a form of governance that anthropologist Elizabeth Povinelli labels "geontopower," an arrangement in which the *bios* of biopower—in this case, black life and black death—is subtended through the management of *geos*, or nonlife.[21] Here, *geos* (for example, water, minerals, petroleum) becomes the active object of intervention and control. Black life, on the other hand, becomes either a disposable means to *geos* or has "promise" by way of future financial extraction. Indeed, post-Reconstruction plantation agriculture, which Harry Haywood analyzed as an imperial formation, produced forms of indebted subjectivity that have become widespread in the neoliberal era.[22] As the petroleum-driven redevelopment of New Orleans turns the region's musical heritage into an investment resource, it ironically highlights the levee's place in popular culture. The levee passes from the material to the metaphorical and back as it circulates (in blues), becomes a cliché (in classic rock), and is then reawakened in a moment of crisis. The repressed collective memory of the levee is exposed by the emergency signs during Katrina and becomes the title of Spike Lee's 2006 documentary, *When the Levees Broke: A Requiem in Four Acts*. If, as Elaine Gan et al. argue, life in the Anthropocene is "haunted" by infrastructure, where the environment takes on a spectral life across diverse histories, blues provides an indispensable *cultural* source of inquiry, creativity, and warning, amidst ongoing endeavors to manage the agential flows of people and matter.[23]

The autobiographical persona of the "blues voice" is celebrated as a literary means for mapping transhistorical experience across social geography. Ellison finds a definitive example of the blues as "an autobiographical chronicle of personal catastrophe expressed lyrically" in Richard Wright.[24] From his plantation birth in Natchez, Mississippi, his move to Chicago, and later travels beyond the US to Bandung, Wright's planetary trajectory anticipates the geography of the blues voice as it extends outward from its "birthplace" in the Delta. His writings exemplify the "dual purposes of autobiographical observation," interrogating self and society in the struggle against racial capitalism and environmental injustice.[25] In the post-flood confluence of disaster, debt peonage, and militarized relief camps, blues gives formal expression to an environmental double consciousness as it is carried to northern cities. From Black Atlantic tributaries to its global post-colonial reception, the selfhood of the blues voice has become articulated to scenes of differential vulnerability in the varied spaces of Anthropocene life—whenever that life is said to have officially begun.

As a genre, blues goes beyond lyrical expressions of the tragicomic in that it serves an epistemic utility for survival. In *Black on Earth: African American Ecoliterary Traditions*, Kimberly Ruffin argues that a blues epistemology may enable us to "think our way out of eco-crisis."[26] She builds on Woods to compile a list of related practices: "transforming painful experience through artistic production"; "gaining insight from the disadvantaged" through "a bottom-to-top perspective on life"; "affirming collective experience with the personal voice"; social critique and "alerting others to crisis"; "offering secular testimony" by portraying "stark" reality;

emphasizing the ability of the self; and a black expression of double consciousness in the face of social contradictions.[27] These elements comprise a practical sensibility that has profoundly influenced twentieth-century culture and cannot be reduced to a single medium. By expanding the "multitude of forms, texts, and archives" that comprise the range of American environmental experience, Ruffin argues that ecocritics can begin "listening to uncelebrated ecological narratives from the past to help us with the present."[28] Turning to overlooked sources may reframe standard histories of environmental experience and provide alternative understandings of human culture in the Anthropocene.

I use the phrase "blues memoir" because it is generic. Framed by the Anthropocene, the micro-taxonomic differences that autobiography studies compile (Smith and Watson list sixty) risk obscuring generative comparisons. To the extent that there can be no "universal" life narrative about climate change or a rift in the Earth system, a more useful approach is to explore how different genres of life writing *articulate*—in the sense of linking—particular lives to longer histories of anthropogenic environmental change.[29] As I argue, in blues memoir, authors' encounters with the blues connect them to a tradition that links their life narratives to broader material memories of planetary production and exchange. The self composed in these texts is profoundly shaped by this history as much as by its own tragicomic sense of agency. Just as memoirs do not attempt a total account of a life, but rather frame the life in terms of a particular dimension, this definition of blues memoir integrates both "testimonial" and "expressive" features that have divided scholarship.[30] Moreover, as blues memoir falls under the category of musical autobiography, it shares the latter's tendency toward collaboration, which further emphasizes the relational, creative lives of its subjects.

Collaboratively written autobiographies underscore the collaboration in the creative lives of these artists. David Ritz is coauthor with B. B. King, and has cowritten autobiographical works with Ray Charles, Aretha Franklin, Janet Jackson, Laila Ali, Willie Nelson, Cornel West, and Tavis Smiley. Through a collaborative process of interviews, editing, and rehearsal arrangements, Ritz shapes the individual voice and consciousness of the life, authoring his or her history into a singular expression. This contrasts with Willie Dixon's coauthor Don Snowden, a music journalist who differentiates his voice from Dixon's through the inclusion of italicized history and multiple-participant interviews. This differs from Arna Bontemps, the poet, novelist, and archivist of the Harlem Renaissance, who edited W. C. Handy's autobiography and likely preferred its original title, *Fight It Out*, on account of its record of racist violence in the South.[31] Here, the amanuensis stands in for hundreds of collaborative musical partners throughout a career that produce a singular voice. The musical dimension of blues memoir challenges notions of transparent meaning associated with legibility. Referring to his childhood stutter, King explains: "Words aren't my friends. Music is. Sounds, notes, rhythms. I talk through music."[32] The *musical* self is defined through performance and iteration, which challenges the unchanging self-identical author associated with writing. In this regard, collaborative life writing draws out distinctions that are commonly overlooked in music, as when the identities of lyricist and performer are conflated. Musical autobiographies necessarily "cross medial and generic boundaries," yet they have their own set of conventions.[33] A common trope, for instance, is the event of the first encounter with music. Citing a "metareferential turn" in Questlove's *Mo' Meta Blues*, scholars suggest that these narrative conventions are becoming self-conscious.[34] Such self-conscious referentiality has long defined blues as a genre that is hyperaware of its transmission, perhaps on account of its uncertain emergence from multiple origins.

The memoirs that follow are each written long after the artists achieved commercial success and influence, and contain retrospective accounts of the industry that shaped their careers. However, this reading focuses on the way these narratives position the artists in relation to the blues itself, and how this relationship discloses alternate understandings of the self and environment within the Anthropocene.

B. B. King, Plantation, and Blues Mothers

Riley B. King, or B. B. King, characterizes *Blues All Around Me* as a "memory of the heart."[35] Cowritten with Ritz, King's *ambient* conception of the blues describes social and environmental relationships as immanent to feeling. These "complicated feelings told in simple stories" characterize his affirmational response to the injustice of the plantation economy as much as his passionate attachments to women and the complications of sexual desire.[36] Born to sharecroppers on a Mississippi cotton plantation, King's relation to the blues was shaped by the women in his family, as well as traveling performers who introduced him to the complex emotions of the music. This sensibility is illustrated in chapters like "Her Words Are Music" and "King Cotton," which link King's encounter with blues to the socioecological history of the plantation economy. The early death of his mother left him to fend for himself, profoundly altering his sense of self and world, yet his response to this loss illustrates similar responses to the material and metaphorical loss of nature in Anthropocene discourse.

In its account of King's early life, *Blues All Around Me* acknowledges the complicated coexistence of suffering and pleasure. That blues offers a narrative ethics, in which contradiction and irony need not resolve in tragedy, is borne out in his childhood encounters with female blues singers.[37] "The scholars loved to praise the 'pure' blues artists or the ones, like Robert Johnson, who died young and represented tragedy," he writes. "It angers me how scholars associate the blues with tragedy." For King, "blues meant hope, excitement, pure emotion."[38] As singers like "Ma" Rainey and Bessie Smith spoke to popular audiences with irreverent songs about "work, jail, prostitution, [and] natural disasters," they addressed topics that compose a "patchwork social history of black Americans during the decades following emancipation."[39] King's account of these influences is joined by his great-grandmother's account of the blues when she had been a slave. "If the master was coming, you might sing a hidden warning to the other field hands," he writes. "That was important for the women because the master could have anything he wanted. If he liked a woman, he could take her sexually. And the women only had two choices: Do what the master demands or kill herself."[40] Amid the sexual and reproductive violence of slavery, blues became an *act* of warning. It had utility: "the blues was about survival."[41] Through this genealogy, King's narrative situates his own emotional formation within the culture of the plantation system. Yet, by refusing the false closure of tragedy, blues affirms alternative trajectories of emotion.

The autobiographical encounter with the blues is narrated in the context of "King Cotton," the name of the nineteenth-century plantation South and its global empire of trade. It is an economic ordering of space, environment, species, and race, which King describes as a "force of nature." It forms an unexpectedly intimate part of his development. "Like the sun and moon and stars above, it surrounded my life and invaded my dreams," he writes. "I saw it, felt it, dealt with it every day in a thousand ways." "Cotton turned me from a boy to a man, testing my energy and giving me what I needed—a means to survive. But I did more than cope with the crop. I actually loved it," he confesses.[42] For King, farm labor provides a satisfaction in that it integrates the temporality of daily life with the rhythms of the natural world. "It was beautiful to live through the seasons, to break the ground in the chill of winter, plant the seeds against the winds of spring, and pick the blossoms in the heat of summer," he writes. "[T]here's a poetry to it, a feeling that I belonged and mattered."[43] Passages like this are accompanied by criticism of Jim Crow and the plantation economy, but serve as examples of the complicated feelings between humans and nature. According to Les Back, King's sense of "elemental vulnerability" in the realm of space and emotion forms the basis of his "planetary humanism" and motivates his prison performances in the early seventies.[44] King challenges the assumption that attachments to the natural world are suspect because people are exploited, confined, or abused by way of the environment. Rather, these attachments become the means for creating alternate temporalities (rhythms), which open dimensions of selfhood capable of feeling beyond present constraints.

In *Blues All Around Me*, King describes the death of his mother and grandmother as the destruction of his world, which leaves him living alone as a child in a sharecropper's cabin. He begins talking to domestic and wild animals. "I felt like they understood me," he writes. And he invents stories inspired by arrowheads from "Indian tribes that once lived on the land."[45] These turns, both to nonhuman animals and to the colonized other whose material traces still inscribe the landscape, enable him to rebuild his world. Later in life, he ritually makes a meal of beans and mashed corn bread on his tour bus as a comforting reminder of the time "after mama died" when he was "doing for" himself.[46] The meal provides a sense of continuity, yet it is a continuity with the self *after* the loss of his mother. The "death of the mother" narrative often marks a simultaneous separation and emergence into "authentic" selfhood, and is instructive for examining Anthropocene narratives through the critical lens of life writing.

In popular Anthropocene narratives, the death of nature as a motherly or harmonious figure is precisely what underwrites the teleological emergence of the *anthropos* as a self-conscious, transcending planetary force.[47] King's narrative illustrates a sense of self that differs from this implied species autonomy. It is a sense of self and world which remains immanent to the loss of his mother. The loss of the transcendent is, in this sense, built into an ethical horizon of meaning. Similarly, in their writings on autobiography, Roland Barthes[48] and Jacques Derrida[49] argue that the mother's death enables the emergence of the *written* self, which postpones the death through writing. The maternal, in their account, is the material precondition for language and "outside" discourse in that "the mother is a metaphor for what is not metaphoric about language."[50] If the maternal is also the genealogical, the blues, like life writing, is a way of reestablishing continuity through performative iteration—even if one's relation to that genealogy is constituted by loss. King's memoir answers the question of how to inherit and inhabit a destroyed world that still contains capacities for joy. The transcendent, maternal metaphor of nature may be gone, yet earthly beings persist.

W. C. Handy as Amanuensis

Born in Alabama, William Christopher Handy was a classically trained musician, professor, and minstrel bandleader, who spent the first decade of the twentieth century in the Delta collecting vernacular African American and Afro-Caribbean folk music, which would later make him the now celebrated figure of Memphis. In his 1941 autobiography, *Father of the Blues*, Handy credits his "fatherly" status to his ability to *transcribe* his listening and publish his songs to wide audiences. For Handy, writing is not only the vehicle for recording, transmitting, and transforming experience into new contexts, but also a means to link his personal biography with the emergence of blues as a commercial art. "There are those who wish me to approach the subject of blues as though this type of music should be shrouded in mystery," Handy writes. "Thousands have heard the material which went into the making of the blues … but they didn't write it down."[51] Yet this humility contrasts with his shrewd understanding of the power of copyright. This is illustrated in Handy's earlier collaboration with Abbe Niles, *Blues: An Anthology*, a popular introduction to the blues told with a biography that sought to counter the dominant discourse.[52] When racialized distinctions between high and low culture still obtain, Handy knowingly risks accusations of primitivism by insisting on the formative influence of folk practices and animals in his music. He employs what Ruffin calls "self-certified biophilia" to affirm positive environmental experience.[53] Rather than view Handy as progenitor, this reading considers how the autobiography positions him as amanuensis to the world of humans and nonhumans often excluded from the domain of culture. As one who listens, receives, records, and arranges dictation, the amanuensis is the underappreciated accompaniment to many collaborative life-writing projects. Despite its title, *Father of the Blues* is as much about Handy's numerous *receptive* encounters with the blues as it is about his inscription of the blues to mass audiences. In an Anthropocene context, Handy's life writing exhibits a blues selfhood that composes itself through relationships to both human and nonhuman others that make it possible.

As Handy recounts the inspiration he receives from daily life, his narrative illustrates Houston A. Baker, Jr.'s "way-station" concept of the blues, which is transmitted "at the junctures [and] intersections" of "discontinuous experience."[54] In a famous passage, Handy wakes up on a train to the surreal sounds of a man playing slide guitar with a knife.[55] Later, he recalls his time with a "shovel brigade" who collectively beat complex tonal rhythms while working in an iron furnace.[56] He turns to childhood experience and memory, describing his reception of a washerwoman's song. First, he reproduces her "moanful voice and unusual tonality" on his violin and then develops it into "three distinct [piano] melodies."[57] Handy's model is Robert Schumann's "Träumerei," or "song without words," from *Kinderszenen*.[58] Handy's appropriation of the

German Romantic brings discontinuous social memory to bear on music. This is similar to the blues sensibility that Cornel West identifies in Theodor Adorno. West distills the German exile's philosophy to one phrase: "the condition of truth [in art] is to allow suffering to speak."[59] Indeed, the year that *Father of the Blues* is published, Adorno composes *Kinderjahr*, a rewriting of Schumann's *Album for the Young*. For Adorno, modern music contains "sedimented content" that "survives what is otherwise forgotten and what is no longer capable of speaking directly."[60] Likewise, Handy's autobiography positions his blues compositional practices as the inscription of memory, in which the received music of labor in fields, shipyards, and prisons rebels against the destruction of experience by the management of bodies and nature.

Nonhumans form an unexpected component of Handy's creative development. Rather than speaking about animals in the abstract, he associates each species with particular kinds of voices, instruments, and expressions of sound, which exist alongside and even respond to the notes of other beings. When he takes up the guitar, for instance, the instrument allows the "riotous carnival" of birds to appear: "perhaps they were not more numerous than they were in the past, but now something sang in me, and I felt drawn to them." "I could tell what the birds in the woodlands and orchards were singing," he writes, and could "visualize the notes on a scale."[61] Handy envisions musical writing that transcribes the multivoiced environment into the harmonic orders of western composition: "Robins carried a warm alto theme. Bobolinks sang contrapuntal melodies. Mockingbirds trilled cadenzas." He writes that "even the bellow of the bull became in my mind a musical note, and in later years I recorded this memory in 'Hooking Cow Blues.'"[62] He describes how the wild and domesticated animals of rural working life "all built up within my consciousness a natural symphony," including the nightly mournful sound of insects: "This was the primitive prelude to my mature melodies now recognized as the blues."[63] By acknowledging this influence, Handy does not imply that the blues is primitive or instinctive; rather, he assertively opens up selfhood to include unacknowledged environmental and nonhuman influences. When a critic suggests these passages be removed, he vehemently demands they remain. "It establishes a very important fact," Handy responds; "the genius doesn't find his music in books and notes but in nature. He hears the music of the stream, the brook, the ocean, and makes notes out of them."[64] Handy's defense advances an agential understanding of the nonhuman environment in relation to human creativity.

This environmental perspective comes not only from a professor of music, but also from an accomplished bandleader who has traveled white and black circuits of American life, and who is intimately familiar with the racial hierarchies encoded in concepts of culture. For instance, the letters between Handy and Niles regarding the 1926 anthology expressly desire to confront both the patronizing gaze of "primitivist historiography" on blues and stereotypes of black childhood.[65] What these passages illustrate is, importantly, an unapologetic affirmation of the natural world at a time when associations with nature were a means of excluding the humanity of oppressed populations. Ruffin terms this move "self-certified biophilia," finding similar environmental validation in the life narratives of former slaves archived by the Federal Writers' Project.[66] Handy confirms Ruffin's thesis that African American literature in the Anthropocene cannot be confined to issues of environmental injustice, but must encompass pleasurable, joyful, and creative engagements with the nonhuman world. Written by the child of slaves, Handy's blues memoir is located in this tradition. However, as a man of the twentieth century, his source of success is the same as his subsequent notoriety as a "blues capitalist." Handy's writing enables him to inscribe his music within the economies of modern sound, but this would not be possible without his autobiographical self-positioning as amanuensis to the music around him.

Willie Dixon's Relational Self

In his 1989 memoir, *I Am the Blues: The Willie Dixon Story*, Willie Dixon identifies with the blues itself. Framed by blues epistemology, Dixon's "I" can be read as an affirmation of collective experience and an invitation for others to join in this empathetic identification. His narrative is a mix of first-person perspective, group interviews, and italicized historical commentary by cowriter Don Snowden. The book's title track proclaims:

> I am the blues
>
> I am the blues
>
> The whole world knows I've been mistreated and misused
>
> …
>
> I'm a thousand generations
>
> of poverty and starvation
>
> I am the dog of the United Nations
>
> I am the blues.[67]

Dixon's collective "I" speaks from the position of the excluded—that is, the exploited within the nations of the world, as well as formerly colonized nations who had begun to exercise their voices by 1969. Yet he goes further than proposing a new abstract universal subject. The relational self of Dixon's memoir is constructed through an understanding of the blues as a critical consciousness of material labor across planetary networks.

Dixon writes with an acute awareness of how labor shapes one's self in the world. Born in Vicksburg, Mississippi, Dixon is best known for the pivotal role he played in the Chicago scene as a musician and producer for Chess Records. His business acumen and labor-mindedness are credited to his work experience. After the big flood, Dixon worked at a government station building experimental "sand and dirt models" for levee and highway construction along the river.[68] Before moving to Chicago, the future blues producer worked in the timber and turpentine industry, the Civilian Conservation Corps, a prison farm (from which he escaped), shipped freight in the Pacific Ocean, and shoveled coal on excursion riverboats between Mississippi and Illinois.[69] Dixon accentuates his worldliness and the worldliness of the blues itself at a time when the genre was associated with regional backwardness. His memoir defines the blues as a referential idiom that by necessity adapts its memory to the changing conditions of work. Citing the line "blues jumped a rabbit and rode it for a solid mile," Dixon argues that it is meaningless to people in cities, who never had to catch a rabbit for dinner.[70] The fact that rural figures, when moved out of their regional context, can lose their capacity to transmit feeling through metaphor means that the blues must "relate to the things of today" if it is to be more than an idiom of references that no longer speak to lived experience.[71]

A similar ethic of blues in transit can be found in the autobiography of one of Dixon's musical collaborators, Buddy Guy. In *When I Left Home: My Story*, cowritten with Ritz, Guy reverses the "standard script" whereby the blues was "blasted out from the plantations where it had been formed" and "dispatched northwards into the accelerated … abject modernity of America's segregated cities."[72] Guy was already electrified when he moved to Chicago from his family's Louisiana cotton field in 1957. "I cottoned to the electricity because it was something I could turn up," Guy writes. "[V]olume did a lot for me."[73] The use of "cotton" as a verb draws on its materiality as an object that sticks to other objects. This idiomatic use of "cotton" as a metaphor is common in the American South and regions of the former British slave trade.[74] To "cotton to the electricity" is to poetically embody the transition from agrarian to industrial economies. Like Dixon, the meaning of the blues is shaped by new attachments in transit but never breaks from its origin. Guy recalls his visit to Gorée Island, Senegal: "one of the places where slaves were sent off to make that terrible Atlantic crossing. Maybe that's where the blues began."[75] His early hit is a rendition of Little Brother Montgomery's "The First Time I Met the Blues," a song about escaping a lynch mob in a forest.[76] When transposed to Guy's Chicago, the space of the forest and the lyrical "you" addressing the mob become a haunting allegory of an existential relation to the self amidst a landscape of urban violence. Published in the wake of the gastro-memoir boom, Guy's narrative is bookended with praise of local organic produce grown among the farms and forests of his suburban Illinois home. At one point a necessity in his life, eating this way is now a marker of middle-class food sensibilities.

Dixon's refusal of agrarian references opens the cultural idiom of the Delta to the global production of late Fordism. He turns blues into a means of relating to others through an environmental consciousness that anticipates the alter-globalization movements a decade later. It is worth quoting at length:

> "If you're writing about cars, you have to think about what the car does … and all the related things from the making of it to the end results of it. If you're going to think deeply enough about it, you gotta think where it comes from and all the various parts. You have to think all the way back to maybe the rubber was made somewhere in Africa on some plantation, guys chopping trees to make the rubber run out. Some guy in some mine somewhere else is mining the copper, another guy's cutting sheep's wool somewhere to make the interior parts of the car and another guy is in a steel mill. The things that make it run come from all different parts of the world."[77]

In these and similar passages, Dixon connects disparate chains of production by invoking the embodied labor of those who comprise the global economy.

At each interval of this network there exists "some guy," a laboring self, located in a particular place shaped by the history of production. His disarmingly informal term qualifies a gendered particular "guy" with a generic "some," but this may also include domestic labor. Dixon imagines his blues in conversation with these various locations. "A guy who ain't directly associated with you is associated with you unconsciously, one way or another," Dixon writes.[78] This "guy" is not the universal Man or humanity, much less the global *anthropos*, the current substitute for the European male subject. Rather, Dixon's "guy" posits a collective subject that is irreducibly plural and materially specific, one who is excluded from the calculations of power yet invited to identify with the address as an imagined alternative to the relations of global capitalism. This passage illustrates how a characteristic feature of blues epistemology in the blues memoir, the bottom-to-top perspective on disadvantaged life, constructs a relational environment of the self. Dixon extends his concern to exploited nonhumans out of whom the very instruments of music are made:

"you might think about somebody cutting trees to make the piano, some poor elephant that dies in Asia for the ivory that's on there, the brass and copper and paint, the nails of the carpenter—it took everybody."[79] Elephants, too, not unlike mules, have a stake in the relational life of the blues.

King, Handy, and Dixon are examples of blues life writing as a source for rethinking environment, culture, and selfhood in the Anthropocene. King emphasizes the power of personal feeling in the ambience of people and place, while Handy emphasizes intersubjective influences at the formal (and informal) level of musical composition. Dixon takes this a step further to *materialize* this compositional imaginary within a global political economy. The self-description as coming from "blues country" casts the Delta neither in the past nor out of time, but as an understory of the planetary present. In the opening pages, Dixon recalls a childhood realization that "if I understood the blues for what they actually are, it would give me a better understanding not only of me and my folks but the rest of the world."[80] To know the blues, he suggests, is to know the relationships that have conditioned identification and selfhood in a global history. As Fanon scholar Lewis Gordon maintains, the blues "reveal the particularity of misconceived universals" and thus "paradoxically transcend their own particularity into a more universalizing practice."[81] While Dixon relates to the blues through this speculative auto-ethnography, the memoir invites the reader to encounter this planetary "country" through particular life stories.

Nothing is as universal today as the market. As musical autobiographies, these texts challenge the presumed link between authenticity and non-market attitudes by accenting the importance of audiences as a vital source of income. Dixon extends the struggle over creativity, ownership, and recognition that we find in both King and Handy to include authorship of blues history through life writing.[82] Recalling "great blues artists that never had a future," Dixon stresses the importance of life writing as both testimony to injustice and creative expression. Referring to the US, he writes, "They didn't write the history of the guy who made it—they wrote the history of the guy who got it."[83] This distinction between agency (production) and control over that agency (ownership) carries Dixon beyond critics who would define agency in terms of mere *affect* to critique an expropriative social relationship inscribed first by material production and then through writing. This critique parallels the split in Anthropocene discourse between the universal activity of the *anthropos*, on the one hand, and the ownership of production, on the other, controlled by successive waves of colonial, plantation, and industrial capitalism whose histories of human difference are under erasure by planetary frames.

The fate of the species identity in the Anthropocene is taken up by the Mississippi-born, white, environmentalist composer John Luther Adams in *Become Ocean*, his 2014 Pulitzer Prize-winning and Grammy Award-winning symphony. "As the polar ice melts and sea level rises," Adams writes , "we humans find ourselves facing the prospect that, once again, we may quite literally *become ocean*."[84] While an intoxicating and beautiful symphony, one remembers that the threat of becoming ocean was first visited upon African captives as a condition of passage into the environment of New World modernity. Moore's concept of "slaveship earth" returns as a metaphor of not only an anthropogenic planet, but also an environment that is intensely social and unevenly distributed. As Freud famously observes, the prospect of oceanic subsumption threatens identity, but it also promises a comforting dissolution of consciousness.[85] Where Adams "invites you to get lost" in a sublime state, unmoored from personal and planetary coordinates, we might question the function of such an aesthetic, and consider how the blues memoir calls attention to the terrain of situated lives and histories.

Handy gives readers a glimpse of a different kind of sublime during the first performance of "Mr. Crump" (later "The Memphis Blues"). The band transforms a crowded Memphis street into a river delta. "Hands went into the air," he writes. "[B]odies swayed like the reeds on the banks of the Congo."[86] Handy goes further than Adams' imagery of humans becoming nature through specific Delta figures of Africa and the US. The Mississippi and Congo are not only parallel "alluvial empires" that link global divisions of race, region, and class.[87] They are joined by deltas in Egypt and India, which were integral to the production and transport of cotton and other goods during the "global reconstruction" of the nineteenth and early twentieth century.[88] Handy attributes his most famous song to cultural techniques "already used" in the Delta by the disproportionately vulnerable: "Negro roustabouts, honky-tonk piano players, wanderers and others of their underprivileged but undaunted class from Missouri to the Gulf."[89] In contrast to the oceanic experience that subsumes difference, swaying reeds suggest individuals in unison, moved by forces of nature yet distinct. Likewise, what is at stake in the blues memoir is not a conception of life in the Anthropocene that is subsumed by environment and geography. Rather, it illustrates how the material environment makes up networks of constraint and opportunity, as well as coercion and creativity, for relational selves.

Conclusion

This reading of Anthropocene lives in the blues memoir hopes to generate further inquiry and to complicate the generalizations that often obscure differential vulnerabilities within this "new" planetary condition. While no argument of this length can encompass the extensive scholarship on fields as diverse as blues, autobiography, and the Anthropocene, blues memoirs, and by extension other forms of blues-inflected writing, offer a starting point for reframing human-environmental relationships in ways that do not reproduce the histories of exclusion they detail. They not only offer examples of *relational* autobiographical selves, but the connection with blues also constructs a particular relational self whose labor and creativity are linked to the racial and environmental histories of the Plantationocene. Against the Anthropocene's double erasure—first, of human inequality by the *anthropos* and, then, of the *anthropos* by geology—life writing is able to refuse the subsumption of the human by drawing into relief the uneven geographies of this process. Returning to Toomer's future-oriented identification with "earth" over a "specific region," his nomadic appeal to universality might instead be understood as a refusal to value some *specific* regions at the expense or destruction of others. At a time when the history of the Anthropocene is being written, it is vital to have alternative figures of planetary relation.

University of Helsinki

Disclosure Statement

No potential conflict of interest was reported by the author.

Notes

1. Gilroy, *Darker Than Blue*, 120.
2. Mizelle, *Backwater Blues*, 51.
3. Eakin, *How Our Lives*, 43.
4. Eakin, *How Our Lives*, 23.
5. Smith and Watson, *Reading Autobiography*, 214.
6. Murray, "The Omni-Americans," 58.
7. See McKittrick, "Plantation Futures"; Haraway, "Anthropocene, Capitalocene, Plantationocene, Chthulucene"; Lewis and Maslin, "Defining the Anthropocene"; Latour et al., "Anthropologists Are Talking"; and Moore et al., "Plantation Legacies."
8. Latour et al., "Anthropologists Are Talking," 6.
9. Latour et al., "Anthropologists Are Talking," 6.
10. Woods, *Development Arrested*, 25.
11. Swanson et al., "Less Than One."
12. Ellison, "Flamenco," 25.
13. Moore, "A Few Words," 3.
14. Hamilton, Defiant Earth, 52.
15. Pulido, "Racism and the Anthropocene," 118.
16. Parrish, *The Flood Year 1927*, 7.
17. Parrish, *The Flood Year 1927*, 7.

18. Parrish, *The Flood Year 1927*, 126.
19. Smith and Watson, *Reading Autobiography*, 215.
20. Mizelle, *Backwater Blues*, 30.
21. Povinelli, *Geontologies*, 30.
22. See Haywood, *Negro Liberation*; Harney and Moten, *The Undercommons*; and Baucom, *Spectres of the Atlantic.*
23. Gan et al., "Haunted Landscapes," 1.
24. Ellison, "Richard Wright's Blues," 129.
25. Mizelle, *Backwater Blues*, 59.
26. Ruffin, *Black on Earth*, 136.
27. Ruffin, *Black on Earth*, 141–142.
28. Ruffin, *Black on Earth*, 26.
29. Grossberg, "On Postmodernism and Articulation."
30. Gussow, *Seems Like Murder Here*, 81.
31. Gussow, *Seems Like Murder Here*, 67.
32. King. *Blues All Around Me*, 1.
33. Stein and Butler, "Musical Autobiographies," 118.
34. Stein and Butler, "Musical Autobiographies," 115.
35. King, *Blues All Around Me*, 2.
36. King, *Blues All Around Me*, 161.
37. King, *Blues All Around Me*, 70.
38. King, *Blues All Around Me*, 23.
39. Davis, *Blues Legacies*, 91.
40. King, *Blues All Around Me*, 9.
41. King, *Blues All Around Me*, 9.
42. King, *Blues All Around Me*, 57.
43. King, *Blues All Around Me*, 57.
44. Back, "How Blue?" 281.
45. King, *Blues All Around Me*, 35.
46. King, *Blues All Around Me*, 313.
47. Hamilton, "Theodicy."
48. Barthes, Camera Lucida, 72.
49. Derrida, The Ear of the Other, 15.
50. Anderson, *Autobiography*, 78.
51. Handy, *Father of the Blues*, 150.
52. Dunkel, "W. C. Handy," 133.
53. Ruffin, "Biophilia on Purpose," 46.
54. Baker, *Blues*, 8.
55. Handy, *Father of the Blues*, 74.
56. Handy, *Father of the Blues*, 139.
57. Handy, *Father of the Blues*, 149.
58. Handy, *Father of the Blues*, 149.
59. West, "Black Intellectuals in America," 84.
60. Adorno, *Philosophy of New Music*, 37.
61. Handy, *Father of the Blues*, 9.
62. Handy, *Father of the Blues*, 14.
63. Handy, *Father of the Blues*, 14.
64. Handy, *Father of the Blues*, ix.
65. Dunkel, "W. C. Handy," 133.
66. Ruffin, "Biophilia on Purpose," 46.

67. Dixon and Snowden, *I Am the Blues*, 1.
68. Dixon and Snowden, *I Am the Blues*, 36.
69. Dixon and Snowden, *I Am the Blues*, 40.
70. Dixon and Snowden, *I Am the Blues*, 3.
71. Dixon and Snowden, *I Am the Blues*, 3.
72. Gilroy, *Darker Than Blue*, 141.
73. Guy, *When I Left Home*, 84.
74. Amner, *American Heritage Dictionary*, 96.
75. Guy, *When I Left Home*, 4.
76. Gussow, *Seems Like Murder Here*, 34.
77. Dixon and Snowden, *I Am the Blues*, 214.
78. Dixon and Snowden, *I Am the Blues*, 214.
79. Dixon and Snowden, *I Am the Blues*, 214.
80. Dixon and Snowden, *I Am the Blues*, 2.
81. Gordon, *What Fanon Said*, 89.
82. Dixon and Snowden, *I Am the Blues*, 38.
83. Dixon and Snowden, *I Am the Blues*, 224.
84. Adams, "Become Ocean."
85. Freud, "Civilization and Its Discontents," 723.
86. Handy, *Father of the Blues*, 100.
87. Woodruff, *American Congo*, 3.
88. Beckert, *Empire of Cotton*, 274–311.
89. Handy, *Father of the Blues*, 99.

Acknowledgments

I would like to thank the Humanities Program at the University of Helsinki for its generous support for this research and the Delta Symposium at Arkansas State University.

Works Cited

Adams, John Luther. "John Luther Adams: Become Ocean." *YouTube*, March 9, 2014. https://youtu.be/yGIEvUOf-JU

Adorno, Theodor. *Philosophy of New Music*. Translated and edited by Robert Hullot-Kentor. Minneapolis, MN: U of Minnesota P, 2006.

Amner, Christine. *The American Heritage Dictionary of Idioms*. New York, NY: Houghton Mifflin Harcourt, 2013.

Anderson, Linda. *Autobiography*. New York, NY: Routledge, 2011.

Back, Les. "How Blue Can You Get? B. B. King, Planetary Humanism, and the Blues behind Bars." *Theory, Culture and Society* 32, no. 7–8 (2015): 274–285.

Barthes, Roland. *Camera Lucida: Reflections on Photography*. Translated by Richard Howard. New York, NY: Hill and Wang, 1981.

Baker, Houston A., Jr. *Blues, Ideology, and Afro-American Literature*. Chicago, IL: U of Chicago P, 1984.

Baucom, Ian. *Spectres of the Atlantic: Finance Capital, Slavery, and the Philosophy of History*. Durham, NC: Duke UP, 2005.

Beckert, Sven. *Empire of Cotton: A Global History*. New York, NY: Vintage, 2014.

Davis, Angela. *Blues Legacies and Black Feminism: Gertrude "Ma" Rainey, Bessie Smith, and Billie Holiday*. New York, NY: Pantheon, 1998.

Derrida, Jacques. *The Ear of the Other*. Translated by Peggy Kamuf and Avital Ronell. Lincoln, NE: University of Nebraska, 1988.

Dixon, Willie, and Don Snowden. *I Am the Blues: The Willie Dixon Story*. New York, NY: Da Capo, 1989.

Dunkel, Mario. "W. C. Handy, Abbe Nils, and (Auto)biographical Positioning in the Whiteman Era." *Popular Music and Society* 38, no. 2 (2015): 122–139.

Eakin, Paul John. *How Our Lives Become Stories: Making Selves*. Ithaca, NY: Cornell UP, 1999.

Ellison, Ralph. "Flamenco." In *The Collected Essays of Ralph Ellison*, edited by John Callahan, 19–26. New York, NY: Modern Library, 2003.

——. "Richard Wright's Blues." In *The Collected Essays of Ralph Ellison*, edited by John Callahan, 128–144. New York, NY: Modern Library, 2003.

Fabre, Michel. *The Unfinished Quest of Richard Wright*. New York, NY: William Morrow, 1973.

Freud, Sigmund. "Civilization and Its Discontents." In *The Freud Reader*, edited by Peter Gay, 722–771. New York, NY: Norton, 1989.

Gan, Elaine, et al. "Haunted Landscapes of the Anthropocene." In *Arts of Living on a Damaged Planet: Ghosts and Monsters of the Anthropocene*, edited by Anna Tsing, et al., 1–16. Minneapolis, MN: U of Minnesota P, 2017.

Gilroy, Paul. *Darker Than Blue: On the Moral Economies of Black Atlantic Culture*. Boston, MA: Harvard UP, 2010.

Gordon, Lewis R. *What Fanon Said: A Philosophical Introduction to His Life and Thought*. New York, MA: Fordham UP, 2015.

Grossberg, Lawrence. "On Postmodernism and Articulation: An Interview with Stuart Hall." *Journal of Communication Inquiry* 10, no. 2 (1986): 45–60.

Gussow, Adam. *Seems Like Murder Here: Southern Violence and the Blues Tradition*. Chicago, IL: U of Chicago P, 2002.

Guy, Buddy, with David Ritz. *When I Left Home: My Story*. New York, NY: Da Capo, 2012.

Hamilton, Clive. *Defiant Earth: The Fate of Humans in the Anthropocene*. London: Polity, 2017.

——. "The Theodicy of the 'Good Anthropocene.'" *Environmental Humanities* 7, no. 1 (2015): 233–238.

Handy, W. C., ed. *Blues: An Anthology, introduction by Abbe Niles*. Bedford, NJ: Applewood, 1926.

——. *Father of the Blues: An Autobiography*. New York, NY: Da Capo, 1941.

Haraway, Donna. "Anthropocene, Capitalocene, Plantationocene, Chthulucene: Making Kin." *Environmental Humanities* 6, no. 1 (2015): 159–165.

Harney, Stefano, and Fred Moten. *The Undercommons: Fugitive Planning and Black Study*. New York, NY: Minor Compositions, 2013.

Haywood, Harry. *Negro Liberation*. New York, NY: International, 1948.

King, B. B., with David Ritz. *Blues All Around Me*. New York, NY: Dey St., 1996.

Latour, Bruno, et al. "Anthropologists Are Talking: About Capitalism, Ecology, and Apocalypse." *Ethnos* 83, no. 3 (2018): 587–593.

Lewis, Simon L., and Mark A. Maslin. "Defining the Anthropocene." *Nature* 519, no. 7542 (2015): 171–180.

McKittrick, Katherine. "Plantation Futures." *Small Axe* 17, no. 3.42 (2013): 1–15.

Mizelle, Richard M., Jr. *Backwater Blues: The Mississippi Flood of 1927 in the African American Imagination*. Minneapolis, MN: U of Minnesota P, 2014.

Moore, Jason W. "A Few Words from the Chair: Slaveship Earth and the World-Historical Imagination in the Age of Climate Crisis." *PEWS NEWS*, May 20, 2018: 1–4. http://asapews.org/pn19-1.pdf

Moore, Sophie Sapp, et al. "Plantation Legacies." *Edge Effects*, January 22, 2019. https://edgeeffects.net/plantation-legacies-plantationocene

Murray, Albert. "The Omni-Americans: Some Alternatives to the Folklore of White Supremacy." In *Collected Essays and Memoirs*, edited by Henry Louis Gates, Jr. and Paul Devlin, 1–190. New York, NY: Library of America, 2016.

Parrish, Susan Scott. *The Flood Year 1927: A Cultural History*. Princeton, NJ: Princeton UP, 2017.

Povinelli, Elizabeth. *Geontologies: A Requiem to Late Liberalism*. Durham, NC: Duke UP, 2016.

Pulido, Laura. "Racism and the Anthropocene." In *Future Remains: A Cabinet of Curiosities for the Anthropocene*, edited by Gregg Mitman, Marco Armiero, and Robert S. Emmett, 116–128. Chicago, IL: U of Chicago P, 2017.

Ruffin, Kimberly N. "Biophilia on Purpose: A Declaration to Become an Ecosocial Citizen." *Intervalia* 3 (2015): 45–48.

——. *Black on Earth: African American Ecoliterary Traditions*. Athens, GA: U of Georgia P, 2010.

Smith, Sidonie, and Julia Watson. *Reading Autobiography: A Guide for Interpreting Life Narratives*. 2nd ed. Minneapolis, MN: U of Minnesota P, 2010.

Stein, Daniel, and Martin Butler. "Musical Autobiographies: An Introduction." *Popular Music and Society* 38, no. 2 (2015): 115–121.

Swanson, Heather Anne, Nils Bubandt, and Anna Tsing. "Less Than One but More Than Many: Anthropocene as Science Fiction and Scholarship-in-the-Making." *Environment and Society* 6, no. 1 (2015): 149–166.

West, Cornel. "Black Intellectuals in America: A Conversation with Cornel West." *Hedgehog Review* 9, no. 1 (2007): 81–91.

Woodruff, Nan Elizabeth. *American Congo: The African American Freedom Struggle in the Delta*. Boston, MA: Harvard UP, 2003.

Woods, Clyde. *Development Arrested: The Blues and Plantation Power in the Mississippi Delta*. New York, NY: Verso, 1998.

Memoir and the End of the Natural World

By Tony Hughes-d'Aeth

ABSTRACT
This essay draws on Dipesh Chakrabarty's essay "The Climate of History: Four Theses" to test the capacity of memoir to bear witness to the Anthropocene. The essay focuses on three texts that feature memoirs of childhood on the wheat frontiers in Canada and Australia—Wallace Stegner's *Wolf Willow* (1962), Barbara York Main's *Twice Trodden Ground* (1971), and Dorothy Hewett's *Wild Card* (1990). As an instrument of colonization and Indigenous dispossession, the impact of wheat was catastrophic, and these memoirs engage with the particular sites and circumstances that shape acts of remembering "wheaten childhoods."

> Yet climate change poses for us a question of a human collectivity, an us, pointing to a figure of the universal that escapes our capacity to experience the world. It is more like a universal that arises from a shared sense of a catastrophe.
>
> — Dipesh Chakrabarty, "The Climate of History: Four Theses"

The question of whether we are now in a new geological epoch—the Anthropocene—is accompanied by a degree of hushed excitement that it seems indecent to notice. Sensible scientists operating according to the codes of their own deep time are sifting the evidence and will one day deliver their *final diagnosis*. In the meantime, like polite hypochondriacs, we maintain an outward sobriety amidst the grave state of affairs, and an inward exhilaration that the worst is about to be confirmed. This melange of indignation and vindication is the obscene edge of an age that is defined by the fact that it is always already *impending*.[1] Regardless of the outcome to this scientific question, it seems that the epistemic consequences of the Anthropocene are already emerging.

This essay both questions and explores the capacity of memoir to bear witness to the event of the Anthropocene. My examples, Wallace

Stegner's *Wolf Willow*, Barbara York Main's *Twice Trodden Ground*, and Dorothy Hewett's *Wild Card*, do not fall within the more self-consciously ecological life writing (eco-memoirs) that has been fruitfully analyzed by eco-critics like Tom Lynch and Jessica White.[2] In a reparative mode, a book like Jerry Wilson's *Waiting for Coyote's Call: An Eco-Memoir from the Missouri River Bluff* records the attempt to live more sustainably with the environment. Stegner, York Main, and Hewett do not document acts of environmental restitution, but recall childhoods spent on wheat frontiers in the midst of the destruction of natural ecosystems to make way for agricultural production. Their works are typified by an admixture of innocence and complicity that derives from the fact that they realize that fate has situated them on the wrong side of history. As adult narrators, they apprehend that they are the children of the perpetrators, and no simple renunciation or recognition can quite wash clean their place in the complex of forces that have both destroyed the natural world and formed their subjectivity. That each of these memoirs takes place on land stolen from Indigenous people interrupts any analysis that presupposes a simple division between pristine wilderness and corrupt "humanity," and leads directly to the question of just who is the "Anthropo" in the Anthropocene. It also questions how settler memoirs recognize the presence of nonhuman agents—in this case, the commodification of wheat—in the materiality of childhood memory.

The study of memoir in the Anthropocene can draw some useful lessons from Dipesh Chakrabarty's influential analysis about the implications that climate change has for history. His essay "The Climate of History: Four Theses" addresses both the formal practice of history as a mode of academic knowledge and the broader question of historicity—the felt sense of being socially in time—in an age where the basic certitudes of "nature" are unraveling under the duress of cascading human interventions. For Chakrabarty, the fact that "humans now wield a geological force" introduces an unprecedented epistemological disruption into the practice of history.[3] As he explains, "[t]he discipline of history exists on the assumption that our past, present, and future are connected by a certain continuity of human experience."[4] But one of the basic historiographical distinctions, he argues, is the separation created by humanism between natural history (which has now become what we call the sciences) and human history. Chakrabarty's first thesis is thus that anthropogenic climate change heralds "the collapse of the age-old humanistic distinction between Natural History and Human History."[5] Although not explicitly spelled out in Chakrabarty's account, the implication is that human history is not by its own devices capable of sustaining the "continuity of human experience."[6] Instead, human history stands on the stage of a deeper, assumed continuity that we have called "nature," and that nevertheless has its own "history." Chakrabarty refines his theses by attempting to reconcile the analysis of global capitalism with the fact that "we have now ourselves become a geological agent disturbing these parametric conditions needed for our own existence."[7] He continues: "At the same time, the story of capital, the contingent history of our falling into the Anthropocene, cannot be denied by recourse to the idea of species, for the Anthropocene would not have been possible, even as a theory, without the history of industrialization. How do we hold the two together as we think the history of the world since the Enlightenment? How do we relate to a universal history of life—to universal thought, that is—while retaining what is of obvious value in our postcolonial suspicion of the universal? The crisis of climate change calls for thinking simultaneously on both registers, to mix together the immiscible chronologies of capital and species history."[8]

Chakrabarty's final propositions thus circle around the diagnosis of a crisis in the subject (agency) of history. Who (what) is to be the *subject* of a planetary history in which humans operate as a species? The nation-state appears manifestly inadequate. Historical materialism seems better equipped to cope with the transnational quality of the contemporary capitalist system, but, for Chakrabarty, the classical Marxist terms of analysis (class, the means of production, surplus value) seem to founder against the reality of climate change. The traumatic epiphany that Chakrabarty's essay describes is characterized by the radical promotion of humanity to the status of a geological force and, at the same time, our radical demotion to the status of species—a status that had been hitherto reserved, in all but the scientific sense, for nonhuman organisms.

This crisis of the historical subject in the Anthropocene is something that would seem to apply a fortiori to the subject of memoir. If even the aggregated elements of capitalism narrativized by historical materialism as "the story of capital, the contingent history of our falling into the Anthropocene," can no longer form a plausible historical subject in the face of the global ecological crisis we are causing, what chance has the mere individual who is the notional subject of a memoir? Memoir seems at best a quaint liberal artifact of an ideology that continues to insist, in the face of the ecological crises erupting in every direction, that history is shaped by individuals. And yet, while the solutions to ecological crises will certainly have to be collective, the form of the memoir might yet offer some guidance through the impasse of historical subjectivity that Chakrabarty exposes. This somewhat surprising claim is based on the fact that what Chakrabarty describes is a crisis in *experience*, and in particular the reflexive problem of *how we experience ourselves*: "Who is the we? We humans never *experience ourselves* as a species. We can only intellectually comprehend or infer the existence of the human species but never *experience it as such*. There could be no phenomenology of us as a species. Even if we were to emotionally identify with a word like 'mankind', we would not know what being a species is, for, in species history, humans are only an instance of the concept species as indeed would be any other life form. But *one never experiences being a concept*."[9] And then again: "We may not *experience ourselves* as a geological agent, but we appear to have become one at the level of the species. And without that knowledge that defies historical understanding there is no making sense of the current crisis that affects us all."[10] It is on the question of *experience* that memoir, and the theories of life writing and psychoanalytic models of subjectivity, might yet provide guidance into the other of ecological being.

As a subgenre of autobiography, memoir is impressionistic and subjective, and a repository for witnesses' accounts of historical events.[11] Memoirs of childhood are particularly susceptible to the vicissitudes of the politics of remembering, and, as we see in recent debates about Hewett's *Wild Card*, the child witness is central in current movements for social justice.[12] By introducing the concept of "wheaten childhoods" to read these settler memoirs, this essay explores how these memoirs engage with the politics of embodiment, memory, and agency in these childhood narratives, where the relationship between the narrating "I" who remembers and the narrated "I" who is recalled is affected by the Anthropocene and the experiences of the self as a geological agent that fall from this. What Chakrabarty's planetary-scale theses seem in the end to assert is that the subject (agent) of history now finds itself *split* (its basic "continuity" severed) by the real of climate change. The historical subject looks into the mirror of nature and does not see its ego-ideal (for example, the metanarrative of its progress) but the abject consequences of its real actions (that is, the externalized by-products of capitalist modernity). How are these memoirs of childhood drawn into Chakrabarty's "climate of history"?

In the remainder of this essay, I would like to tease out the homology between the split in the historical subject caused by climate change[13] and the seemingly more mundane split that occurs in memoirs between the *subject of action*, the narrated "I," who is acting out the events in the historical time of memoir, and the *reminiscing subject*, the narrating "I," who is inscribing their recollections in the now of the writing moment. While this is a structural split visible in all memoir, even the most immediate reportage, the split between the acting and reminiscing subjects is particularly accentuated when memoirs restage the events of childhood. For Stegner, Hewett, and York Main, whose childhoods were shaped in the interwar years on settler-colonial wheat frontiers in Canada and Australia, the split became—as we will see—almost unbearable.

The Empire of Wheat

Each of the three writers considered in this essay grew up on wheat farms in the former British Empire. Stegner, the Pulitzer Prize-winning novelist, essayist, and memoirist, grew up in the town and adjoining district of Eastend in far southwestern Saskatchewan. His family had migrated to Eastend in 1914 (via Montana and Iowa) to take up land being released for wheat farming. Hewett, one of the leading Australian writers of the postwar period, grew up in the district of Wickepin in the Western Australian wheatbelt on land selected by her maternal grandparents in 1912. York Main, like Hewett, grew up on a farm in the Western Australian wheatbelt, before going on to become an eminent zoologist specializing in trapdoor spiders, as well as a creative writer of note. York Main's father and uncle had emigrated from Yorkshire to Western Australia in 1909 to take up land at Bungulla near Yorkrakine Rock. I have used the words "selected" and "take up" to refer to a process whereby land was issued by governments to "settlers" (private individuals or families) for a nominal price and on the condition that it be converted from its "natural" state into productive farmland. These words are euphemisms that erase the dispossession of Indigenous peoples that proceeded apace on these wheat frontiers. In both Australia and Canada, the legislative instruments for this kind of land issuance were modeled on the 1862 Homestead Act, which paved the way for the close settlement of the Great Plains of the US. The land rush that ensued rolled across the North American continent for half a century but was drawing to its end by the early years of the twentieth century, with places like Eastend in Saskatchewan representing the final throes of this massive repurposing of the prairies for cropping.

During the period of settler expansion, wheat was shipped throughout the British colonies and was often one of the first plants grown when a colony was formed. Wheat was a vital agent in settler colonialism, although as nonhuman matter its commodification and instrumentalization frequently remains unobserved. The materiality of wheat and its agential status on these frontiers becomes visible in recent theorizations of posthumanist approaches to life-narrative scholarship.[14] British colonies in both Canada and Australia originally grew British wheat varieties but, as the colonies expanded, the wheat found itself in conditions much drier (Australia) or colder (Canada) than it had been used to. The introduction of a Scottish variety, which came to be known as "Red Fife" or "Canadian Fife," in 1842 allowed for a dramatic expansion in wheat acreage in Canada. In this sense, it can be said to have *caused* the initial agricultural colonization of the prairies, laying railways and sprouting towns in its wake.[15] But Red Fife was also relatively late-maturing and susceptible to frosts, which are a feature of autumns in the Canadian prairies. In the 1880s and 1890s, the Canadian Experimental Farms Service began interbreeding Red Fife with an early maturing Indian variety, "Hard Red Calcutta," and the result was dubbed "Marquis." Marquis quickly became the wheat of choice in western Canada and is almost certainly the variety that the Stegners would have planted at Eastend. By 1918, 20 million acres of Marquis were planted from northern Saskatchewan to Nebraska, with a worth of around 500 million dollars. It made Canada the largest wheat-exporting nation in the world and fed the Allied war effort in World War I. Meanwhile, in Australia, problems with black-stem rust (a debilitating parasite) had driven wheat-growing into drier areas away from the coast, and new wheat strains were needed to cope with these arid conditions. William Farrer led efforts to develop rust-resistant, drought-tolerant wheat and, like the Canadians, he turned to early maturing Indian varieties.[16] Farrer's "Federation" wheat was named in honor of the Federation of Australia in 1901 and combined Indian ("Etawah"), Canadian ("Red Fife"), and European ("Purple Straw") varieties. Federation wheat and its descendants quickly spread throughout the Australian colonies, adding to the yield in existing areas but also driving wheat-growing into vast areas of the inland that had hitherto only been used in the pastoral industry. Federation also impelled the spectacular growth of the Western Australian wheatbelt, which drew Hewett's family onto the land.

In Australia, the wheat frontiers proceeded more spasmodically than in North America. South Australia was an early wheatbelt, settled in the 1860s and 1870s to service the booming Victorian colony. Western Australia was, like southwest Saskatchewan, a relatively late settler-colonial land frontier. It attained its basic shape in the first thirty years of the twentieth century when a confluence of demographic, fiscal, and scientific developments made viable the conversion of the southwestern corner of the Australian continent to cropland. While the Great Depression arrested land clearing in the 1930s and 1940s, favorable postwar commodity prices and advances in agricultural science led to a second wave of clearing. In two 30-year periods—from 1900 to 1930 and from 1950 to 1980—roughly 0.25 million square kilometers were cleared for intensive agricultural production, mainly wheat and wool. This is equivalent to the area of Britain or Minnesota. In this respect, the creation of the Western Australian wheatbelt during the course of the twentieth century is a significant socioecological event in its own right. But it was also part of a much larger process whereby the world's temperate woodlands and grasslands were converted to intensive agriculture by a mixture of technological and economic forces.

By the early nineteenth century, the Industrial Revolution and European imperialism had intertwined to produce what James Belich called the "settler revolution."[17] In a succession of theaters characterized by various "rushes" (gold, timber, land), the settler revolution led to the explosive settlement of the Americas, Australasia, and southern Africa. The consequences for Indigenous peoples were typically devastating, as their lands were expropriated to make way for settler farms. But the process also produced far-reaching ecological alterations as endemic ecosystems were replaced by introduced agricultural monocultures. During the nineteenth and twentieth centuries, a large portion of the earth's surface—from Siberia to the High Veldt, from Manitoba to the Argentine Pampas, from the Darling Downs of Queensland to the Canterbury Plains of New Zealand's South Island—was transformed with an unprecedented velocity. With its planetary scale, and its far-reaching deterritorialization of natural ecosystems, it is clear that the settler revolution, as well as being devastating for the sovereign Indigenous people in these lands, was a *significant vector of the Anthropocene.* What this means is that the memoirs which Stegner, Hewett, and York Main wrote of their experiences growing up on wheat farms selected through the mechanisms of settler colonialism (that is, the respective land legislation of governments in settler states like Canada, the US, and Australia) also bear testimony to the constitutive events of the Anthropocene.

Wheaten Childhoods

Stegner's *Wolf Willow* was published in 1962 when the author was fifty-three years old, while York Main's *Twice Trodden Ground* was published in 1971 when she was forty-two years old. Hewett's *Wild Card* was published in 1990 when Hewett was sixty-seven years old. While each book functions as a memoir of early life, the books are rather different in their mood and scope. Indeed, while I am characterizing all three books as autobiographical, only Hewett's is an autobiography in the conventional sense. Her book, subtitled *An Autobiography, 1923–1958,* covers the first thirty-five years of her life, including the first eleven spent on "Lambton Downs," her family's wheatbelt farm. *Wolf Willow* is an intriguing amalgam of fiction, history, and memoir. Stegner uses the "madeleine" of the scent of the wolf-willow blossom to open an interface between his formative memories and a later historical consciousness of the deep forces that caused his family to be farming wheat, albeit with little success, in Saskatchewan between the years 1914 and 1920. York Main's *Twice Trodden Ground* consists of seventeen short semi-fictional stories, illustrated in pen and ink by the author, which depict encounters with the countryside of the wheatbelt. Like Stegner, it seems that, for York Main, the story she wants to tell cannot be made to sit in either of the conventionally separate domains of fiction and nonfiction, but requires a slightly ungainly co-opting of both. The vignettes in *Twice Trodden Ground* often use the first person of memoir, but they do not tether this narrator directly to "Barbara York Main" through any direct internal textual evidence. Sometimes, the characters resemble the author and her young family, but at other times they seem to be channeling the anecdotes of others. Several of the stories stage fictionalized encounters between representatives of different farming generations, in which the younger generation asks searching questions of the older generation about the world they are inheriting. What draws all three books into an uncanny orbit is the fact that the authors became absorbed—at more or less the same time and on opposite sides of the world—into the deep ideological currents of the settler complex during their early infancy. Hewett and York Main lived on their farms from birth, Stegner from age five. Both Hewett and Stegner left their farms for good at the age of eleven, and while York Main remained connected to her farm, she began a life away from it when she started boarding for her secondary school and then moved to Perth to pursue her university degree. Yet, for each of these authors, these early years in their life provided the touchstone of their imaginative lives and determined in one way or another a good part of their celebrated writing careers.

Although the end product in each case was primarily wheat, the kind of farming that took place in southwest Saskatchewan and southwest Australia in the opening decades of the twentieth century was subtly different. The Stegners' half section of 320 acres was, for instance, about a tenth of the size of the three thousand acres held by Hewett's parents and grandparents. The Hewetts' farm was itself notably large: the standard wheatbelt farm at this time, such as that selected by York Main's father, was one thousand acres. That said, when Stegner recalls the twenty-five bushels per acre his family farm yielded, this was about 250% higher than wheat yields in Western Australia in the 1920s and 1930s, even with the application of substantial quantities of superphosphate. In other words, and through no mere coincidence, a 320-acre half-section farm in Canada produced about the same quantity of wheat as a standard thousand-acre Western Australian wheatbelt farm in the interwar years. The significantly lower yield, and thus the larger acreages given to each settler, meant that the Western Australian wheatbelt sustained a population density much lower than Canada. The impact of wheat yields on childhood experience was significant: for example, this was one of the reasons why both Hewett and York Main were schooled by correspondence. In many cases, it was difficult to aggregate sufficient students to create even a single-teacher school. By contrast, the Stegners spent summers on their farm and winters in town. This kind of dual domicile was, by and large, not a settlement pattern seen in Australia.

Deep Maps

We might understand the capacity of memoir to offer testimony and bear witness to the Anthropocene by drawing on the concept of the "deep map" that has emerged in the study of place-based writing in the Great Plains. The term derives from William Least Heat-Moon's book *PrairyErth: A Deep Map*, a study of the complicated history and geography of Chase County in Kansas, which contains one of the few vestiges of the tall-grass prairie that once spread across the entirety of the eastern Great Plains. In *Deep Map Country: Literary Cartography of the Great Plains*, Susan Naramore Maher extended the concept of the "deep map" to books like Stegner's *Wolf Willow*. For Least Heat-Moon and Maher, the "deep map" draws on the concept of "deep time" (of "natural" and "pre-" history) to perform a kind of textual decolonization. If Jefferson's 1785 Land Ordinance and the 1862 Homestead Act had parceled out the lands west of Pennsylvania into 640-acre *sections* (a section is a square mile), eradicating Native peoples and ignoring environmental determinants, then the "deep map" is an attempt to restore complexity to the worlds abstracted by this colonization. The complexity that deep-mapping brings back into the picture (its "depth," in other words) encompasses all the things that were made to disappear both in the colonial process and the celebratory metanarrative that emerged to validate it. These vanquished elements included the pre-European dispensation of people and ecology, the violence and perfidy that accompanied the acquisition of land, and all the messy interactions and sequelae that have trailed in the wake of the frontier. In short, a deep map is a kind of anti-map: instead of clearing a surface so as to spatialize in two dimensions a fixed set of geographical variables, the deep map ruptures the surface to reveal a complex interweaving of non-assimilable layers. In this respect, *Wolf Willow*'s genre-defying palimpsest turns out to be, for Maher at least, paradigmatic of the deep map.

Within the terms established in the North American context, neither *Wild Card* nor *Twice Trodden Ground* quite amounts to the "deep map" that is generated in Stegner's *Wolf Willow* or Least Heat-Moon's *PrairyErth*. In the case of Hewett's *Wild Card*, for instance, there is a notable absence of the kind of sweeping historical vision that characterizes the books by Stegner and Least Heat-Moon. There is an insistent exceptionalism in Hewett's account of her family's life story that situates it outside and beyond the story of the wheatbelt, rather than symptomatic of it. The result is that the socioecological event of the wheatbelt is telescoped through the intensities of her family dynamics. There is also an urgency and passion, as well as an intense bitterness, in Hewett's *Wild Card* that is largely absent from the more circumspect narratives in *Wolf Willow* and *PrairyErth*. The deep-map stories of Stegner and Least Heat-Moon, that Susan Maher explores in Deep Map Country, find solace in the "deep time" of prehistory. The traumas they describe—genocide, ecocide, dust-bowl depopulation—are mitigated by the multiplication of parallel and overlapping stories. But in Hewett's autobiography, these kinds of historical and material determinants are fairly muted, and one gains the impression that the farm has been brought into being by the sheer force of her maternal grandmother's will. While Stegner's book draws repeatedly out from the loneliness and wonder of his childhood to the epic history and geography of the Great Plains, Hewett's book remains resolutely centered in the tortured dynamics of her family life—the bleak melodrama of her parents' and her maternal grandparents' relationships, told with unsparing frankness, and the scandal of her own burgeoning sexuality.

Compared with Hewett's memoir, York Main's *Twice Trodden Ground* fits more readily into this concept of a deep map, particularly if one joins this book to her earlier natural history, *Between Wodjil and Tor*, which depicts a year in the life of the flora and fauna of the interlocking ecosystems that surround the granite outcrop known as Yorkrakine Rock. The two books, *Between Wodjil and Tor* and *Twice Trodden Ground*, work in many respects as companion volumes inasmuch as they depict the same matrix of land in the immediate vicinity of York Main's family farm. In Chakrabarty's terms, the two books keep alive a distinction that he now sees collapsing between natural history (*Between Wodjil and Tor*) and human history (*Twice Trodden Ground*). The title of the latter book refers to the fact that the land of the wheatbelt had first been "trodden" by the bare feet of Indigenous Australians and then, much more recently, had felt the far heavier tread of European boots. For York Main, these first people have departed at some point prior to her arrival and are never to return.[18] Aboriginal people only exist as a trace: "They left only signs and artifacts on camping grounds and around waterholes—apart from that the countryside was left in its natural state."[19] This view of a "nomadic" Indigenous population that evaporated as the Europeans arrived obviously bypasses the fact of their concerted and violent displacement from their ancestral lands, but it was a view that it was still possible for a sensitive intelligence to hold in 1971.

Yet, despite the fact that York Main relies rhetorically on the myth of the evaporating indigene, *Twice Trodden Ground* is not a triumphalist account of progress. In this, her book represents a significant shift away from the shire histories of the wheatbelt that had started to appear in the decade or so prior to *Twice Trodden Ground* and toward a process of "deep mapping" in which complexities are explored and the profound losses caused by settlement are not immediately transposed as gains ("progress") in the settler-colonial metanarrative. Indeed, York Main's book is partly written as an answer to these shire histories, which were commissioned by wheatbelt shires to provide a celebratory history of their triumph over natural (drought, flood, fire, pestilence) and human adversity (war, economic challenges). York Main sets her own reflections apart from these shire histories, acknowledging that they are directed to a different register of effects: "The following essays do not amount to a shire history. They hint rather at the effect of place on human emotions as much as the effect of man on place. Or is it only my own response to locality that I present? I do not think so. The potency of place—of landscape—is sometimes of lifelong recurrence, even though the response to its influence may be unformed, unvoiced."[20] Despite this typically modest characterization, it quickly becomes clear that York Main's work is not just a slightly more personal version of the story given in the shire histories, but a radical counter-history of the wheatbelt—one which opens it up to a searching and far-reaching critique. Where the theme of the shire histories is achievement, what has been gained, the motif of *Twice Trodden Ground* is decimation, what has been lost.

York Main's "revisits" to the wheatbelt as an adult (mother, zoologist) become both the analogy and the experiential means of seeing the wheatbelt as doubled—or split—by the effects of its agricultural colonization. The word York Main uses to describe this uncanny doubling is "superimposed": "I spent my early childhood in an environment which to a large extent still showed the imprint only of the first human tread. It was a landscape primarily of virgin bush. On my subsequent revisits the countryside has exhibited the full impact of the second tread—all the modifications wrought by settlement superimposed on the slight signs of the earlier tread of nomadism. Throughout the following sketches this double image is implicit: the broader one of a landscape touched by two cultures and the personal one of this same landscape, first embracing my early life, and later re-entered by my adult self."[21] The wish that the stories evince is that this gap might somehow be closed, that the childhood world be recovered, that the bush which has now vanished be restored or, at the very least, that these losses find some adequate form of symbolization. Like Chakrabarty, York Main is trying to locate a stable (that is, continuous) ontological ground in which to situate memory, life, and history: "In countering despair I have searched for a *continuity* in such change, thereby attempting a reconciliation."[22]

The Tense of the Anthropocene

It seems somewhat surprising that the epochal challenge of the Anthropocene might be reduced to matters of grammar, but the historiographical problems that Chakrabarty diagnoses were partially laid bare by the difficulty posed by the collective pronoun: that is, an "us" to designate "a figure of the universal that escapes our capacity to experience the world."[23] But which comes first, grammar or experience? In the humanism of our common sense, grammar is just a set of rules we use to express what we imagine to be an a priori reality that we apprehend empirically. But what if we apprehend grammatically? And what if grammar was the mechanism that produced us as subjects? Along with pronouns, we might also consider the foundational role played by tense in the basic situation of the subject in time. In the typical subject-verb-object structure of English sentences, for instance, the pivoting verb has a tense which also often acts as a proxy for causation in the sense that we assume the present to be caused by (conditional on) the past.

It is in this respect that Hewett's *Wild Card* offers something that the more circumspect accounts of Stegner and York Main do not, which is a collapsing of tense that seems somehow to capture the temporal incommensurability of the Anthropocene—or, to put this back into Chakrabarty's terms, a temporality that has the capacity to house the paradoxical subject of history based on "a universal that arises from a shared sense of a catastrophe."[24] What is striking about Hewett's memoir is that it does not seem to be afforded the protection of a future in which the things described have *already happened*. Instead, her reflections remain raw and explosive, as if they are happening still for the first time:

> A ten-year-old, carrying the billy of tea in one hand and trying to keep her brown lisle stockings up with the other, I am taking the afternoon smoko to "the man" ploughing the hundred acres across the creek.
>
> I've never worn stockings before and I've forgotten the garters (or perhaps I've never had any). They concertina around my knees and fall to my ankles. The billy bumps, scalding, against my thigh.
>
> I am very conscious of the handsome Italian farm labourer watching my stump-jump progress, hiding his smile with his hand. He is kind to me, the first of many men who will find my aspirations forgivable, even charming.
>
> But they are neither forgivable nor charming.[25]

In fact, Hewett's memoir tends to be written in the present continuous tense—"I am taking" as opposed to "I take" or "I took"—where the events seem to have an ongoing quality. This tense is visible in many parts of *Wolf Willow* but is continually intersected, and thereby framed, by the broader historical narrative. In Hewett, the enchantment remains intact, so that the fairy-tale world of her childhood seems to pass almost without interruption into the traumatic events of her adulthood—her suicide attempts, abortions, broken relationships, and the conversion to communism that turned her away from writing for more than a decade. The spell seems never to be broken. Instead of historical perspective, the narrative introduces a melancholy distance between the subject and her life that seems sourced in the deepest springs of familial resentment.[26]

But perhaps the gap between Stegner and Hewett is not as great as it first seems. One must lay some significant weight, for instance, on the revolution in public taste that occurs in relation to the disclosure of a family's dirty linen between 1962 and 1990. One need only think of the publication of Robert Lowell's *Life Studies* in 1959 to recall the transformation of poetic sensibility that made the discussion of shameful thoughts, states, and actions (infidelity, promiscuity, mental illness, suicide) de rigueur where once they had been considered unprintable except through a thick veil of euphemism. Then there is the crucial matter of gender and the sense in which a man has the opportunity to become disembodied in ways never quite open to a woman. This is a more difficult point to sustain, of course, but reading Hewett next to Stegner is a powerful reminder of what sexual difference decides at the level of our intimate subjectivity.

Where Stegner and Hewett seem most united is in the way that they recollect their childhood. Despite the differences in shade that I have described, each vividly recreates the mixture of wonder and bewilderment that attaches to that phase in which the human subject assumes its symbolic coordinates. On the one hand, there is the fully libidinal dimension of childhood subjectivity: Stegner describes his younger self as a "sensuous little savage."[27] On the other hand, there is the evidence, too, of the way in which each of these children in their frontier wheat farms at the edges of empires was what you might call a fully interpellated subject. That is to say, they emerge as rather pure representatives of the ideology of these late chapters in settler-colonial history—something their adult selves find curious and appalling. As children incorporated in the pioneer mission, they took the imperatives of this project to be, as it were, the laws of the universe. As Stegner wrote elsewhere: "Eastend was a failure place. … But it held our atomic family together for six years … and it imprinted me, indelibly, with the perceptions, images, memories, behavior codes, and attitudes that have controlled my mind and life and writing ever since."[28] In *Wild Card*, Hewett writes wryly: "As far as the eye can see stretch our three thousand acres. We are amongst the richest farmers in the district. We *know* who we are."[29] It is at this subjective level, rather than through the restoration of blighted ecological and cultural dynamics, that Hewett, York Main, and Stegner share a particular relationship of depth. The depth is not that described by Maher in her delineation of the "deep map." If there is a shared mapping at work in these three writers, it is in mapping the human heart and the particular cartographies of settler desire. All three were forged in that particular and powerful current of social desire which is sometimes called "land hunger" and was such a crucial ingredient in the agricultural colonization of the globe in the nineteenth and twentieth centuries. The deep maps that Stegner, Hewett, and York Main draw in their memoirs are ones that capture the generational effects of settler-colonial land hunger.

Returning to Nothing

What becomes clear in the memoirs of all three writers is that the pioneer mythology that spins its story around the vicissitudes of land hunger has a traumatic center. The center of trauma is none other than the homestead that served as both the site of their childhood and the hearth of pioneering experience. Although their books differ in many formal respects, Stegner, Hewett, and York Main each narrate an event that is startlingly consistent: the return in their adulthood to the home of their childhood. This event, part real, part mythical, is the punctum of each text and marks them as "postpioneering" narratives, to borrow a term from Robert Thacker.[30] For the student of memoir, what this moment enacts in the gesture of returning to the childhood home is the memoirist (and their reader) coming face-to-face with the constitutive split of subjectivity, a split that may just provide us with the mechanism for making sense of a nature that has, in the Anthropocene, become unnatural.

For Stegner, Hewett, and York Main, the return to their childhood origins as an adult is utterly horrific. Indeed, Stegner simply cannot bring himself to visit the site of his family's farm, admitting frankly, "I am afraid to."[31] Returning to the ruins of this farm threatens to undo his basic sense of identity: "I can imagine actually finding the flat on which our house stood … locating the place and standing in it ringed by emptiness and silence, while the wind fingered my face and whispered to itself like an old blind woman, and a burrowing owl, flustered by the unfamiliar visitor, bowed from the dirt mound of its doorstep, saying, 'Who? Who?'" He goes on: "I do not want that. I don't want to find, as I know I will if I go down there, that we have vanished without a trace like a boat sunk in mid-ocean. I don't want our shack to be gone, as I know it is; I would not enjoy hunting the ground around it for broken crockery and rusty nails and bits of glass. I don't want to know that our protective pasture fence has been pulled down to let the prairie in, or that our field, which stopped at the Line and so defined a sort of identity and difference, now flows southward into Montana without a break as restored grass and burnouts. Once, standing alone under the bell-jar sky gave me the strongest feeling of personal singularity I shall ever have. That was because it was all new, we were taking hold of it to make it ours. But to return hunting relics, to go down there armed only with memory and find every trace of our passage wiped away—that would be to reduce my family, myself, the hard effort of years, to solipsism, to make us as fictive as a dream."[32] What Stegner is articulating here is captured by the recent coinage "solastalgia."[33] Whereas in nostalgia (the word on which the neologism plays) the subject is caught in the powerful ambivalence of their former life and times, in solastalgia the subject is overwhelmed by the devastating realization that the natural world (what Chakrabarty calls "the parametric conditions" of life) is irrevocably gone.[34] The unbearable thought is not simply that things change, but something like: "The world in which 'I' was made to exist no longer exists, so in what sense is it possible that 'I' am still here? And, in point of fact, where is 'here' now that 'there' is gone?" This is the solipsistic nightmare, in which the ground begins to disappear beneath the ego's feet, which Stegner attempts to circumvent by electing not to see the ruins of his childhood farm.

What is implicit in such moments of solastalgic dissolution is the psychoanalytic process known as identification. While the concept of "identifying" with something has passed into general parlance, where it is often collapsed with the concept of empathy, the nuts and bolts of the process are often overlooked.[35] The central gesture in identification is an unconscious assertion: "That's me!" But, as Jacques Lacan emphasized in his formulation of the mirror stage, this equivalence transposes the supposed unity of the *object of contemplation* into an imagined unity in the *contemplating subject*.[36] If this sounds similar to the split between acting and narrating subjects in memoir (and narrative more generally), then that is hardly accidental.[37] In the case of Stegner's account of his non-visit to his farm, the missing premise is simply the identificatory assertion that this farm, which no longer exists except in the poignant residue of its detritus, is *me*. But here we must introduce a crucial modification to the now popularly accepted idea of identification, which is to distinguish between two levels of identification, both of which are usually unconscious. The first is the identification of the ego with its ego-ideal, the image of its best self. The second is an identification made by the subject of the unconscious with its own abject status outside the symbolic coordinates of reality. Stegner's (non-)return to his childhood home is a way of capturing the duality at the heart of identification, in which the moment and place which "gave me the strongest feeling of personal singularity I shall ever have" was also at the same time completely invisible and incapable of inscribing itself into the world of reality except as debris ("broken crockery and rusty nails and bits of glass").

In the case of Hewett, she does indeed return to her family's farm that was known to them as Lambton Downs. The coda of *Wild Card* features her revisiting the ruined farmhouse with her sister, both now in their sixties:

> We drove on across the railway line to the boundary of Lambton Downs. On the other side of the iron gates was a hose running hopelessly into a sheep trough, with a hole in it. Nobody came this way now. The roads were grown over, the gates wired up. Desolate, treeless except for the hardy York gums and jams, the crossing over the creek was barely distinguishable, but lying in the hollow were a scatter of corrugated-iron sheds and a single brick chimney.
>
> "Oh God!" we cried. "Oh God! Is that the house?"
>
> Two almonds, a few figs and one quince had survived. No she-oaks, no wattles, no tea-tree, no paperbarks, no bottlebrush, no salmon gums, no stables or sheds or post and rail sheep yards, only the concrete dip left like a scar in the home paddock, littered with iron and rusty machinery. Day's timber all gone—a denuded landscape.[38]

Here we can note a basic difference between Stegner and Hewett. For Stegner, the encounter with the reality of loss—that the farm no longer existed—threatened to undo his relationship to reality as such, a collapse into a nightmarish solipsism. For Hewett, though, there is a perverse security in the fact that while reality has been a derelict curator of her past, the crucial libidinal investments have yet been preserved in the fictional worlds that found expression in her literary works. But once again, perhaps the difference is more apparent than actual, as we recall the crucial role that the fictional episodes play in *Wolf Willow*. While these fictional stories awkwardly complicate the generic location of Stegner's book, they also serve to secure the subjective location of the settler in something other than the traces of memory. Yet, in each case, the return is characterized by the same acute sense of solastalgic identification, in which the disappearance of the home is the proxy for the irrevocable disappearance of childhood.

In York Main's case, what is significant is that she sees the "vanishment" of her home in ecological terms. She calls the natural world, which her later career as a zoologist would minutely document, the "substrate" of her own existence. But, as she asks in the preface to *Twice Trodden Ground*, "what … [happens] when one returns to find on the original substrate only fragile remnants of the former landscape!"[39] The solastalgic encounter with her home is intriguing in York Main's case because, unlike Stegner and Hewett, the house is still there and now accommodates her brother's family. And yet, the fact that it is no less gone than the ruined homesteads of Hewett and Stegner exposes the fundamentally imaginary quality of what we call home. In the story "Trespassing," the narrator returns with her children to her childhood home only to experience it as a kind of "double photographic exposure," in which the current state of the home is overlaid by her primordial memories of it. Details from the home bring to her mind trails of remembered actions and moments. In the backyard, now a tamarisk grove, she sees the chain swing that once lifted her "out of the mundane world and into the sky."[40] A deep ambivalence sets in that is quite distinct from conventional nostalgia, and under whose influence the narrator comes to feel like a "ghost of the dispossessed": "It is a painful paradox that long after one is alienated, even if voluntarily, from the site of one's beginnings, a sense of proprietorship remains. But reality cannot recognize such proprietorship—*it is trespassing*! And in my trespassing I had unwittingly discovered that other self, and in my discovery, it finally withered. I fled that house, for who has the courage to look into a tomb?"[41] It is an evocative and complex moment. The thrust is straightforward enough: one cannot live in the past, *trespass* on the past. The drama turns on the meeting of two selves: the seeking, present-day narrating "I" and this "other self," later called "the child," who is anterior and hidden until the act of trespass.[42] It is the trespass itself that makes this other self visible and prompts their meeting, a radical encounter in which one flees and the other withers.

What are we to make of this? And what are we to make of the lingering question "who has the courage to look into a tomb?" The key for me lies in the metaphor of *trespass*, which expresses as a juridical verb the foundational element of property—namely, a state sanction to exclude others from a thing. In this sense, *trespass* imports the prohibitive force that taboos hold in traditional society and stitches them into the subject. This poignant interior drama is acted out for York Main at a socioecological level in the story of the Western Australian wheatbelt, and she sees in each *scene* an answering desolation, which she describes in the succeeding paragraph of "Trespassing": "For the child had gone long ago, with the vanishment of scrub and trees. Now it was only in a mirage of fancy that this landscape lived again, when in the falsity of reminiscence thickets of tamma and calothamnus and sheoke, hakea and timber, spread out over the swell and hollow of the rolling countryside. Only a little remains, small primeval islands in the great surround of open paddocks. It is to these last relics of unchanged reality, to these islands of yesterday, that one desperately lays claim."[43] York Main thus finds in her own sense of loss felt toward the missing bush the echo of a loss felt in her childhood—a loss then re-experienced as the passage *out of* childhood. Childhood becomes the storied location of that primal loss, but the linkage is more than a mere psychological projection. Whether Barbara, the narrating "I," had had a happier childhood would not change the ecological destruction that paved the way for the wheatbelt. The loss of habitat and the loss of "the child" are both registrations in the symbolic. They are holes that we circle incessantly and around which we build the fences that constitute culture and mark our entry into the adult world. It is, indeed, the experience of losing the child that makes feasible the discernment of shifts in the real, shifts that, once temporalized, appear before us—in York Main's beautiful phrase—as "islands of yesterday." It is also this experience that holds the key to a movement past the tomb of primal loss and into a living present and future. Perhaps a traumatic encounter with loss is a necessary wound, and our adulthood forms only in the scar tissue that it instigates. In any event, traumatic loss is often the affective state yielded by modernity's acceleration, by its combustion of the past to fuel an impending unseeable futurity.

Conclusions

This essay questions the capacity of memoir to bear witness to the Anthropocene. These case studies recall the final phases of the settler revolution, which played such a significant role in transforming the temperate latitudes of the planet by converting, often on a continental scale, endemic grasslands and forests into cash-crop grain monocultures. On reading an early draft of this essay, a historian friend of mine was struck by the fact that the unit cost of wheat as a global commodity exactly determined the size of the farms in the Western Australian wheatbelt (1,000 acres) and southwest Saskatchewan (320 acres) in opposite corners of the world and in climates that seemed so vastly disparate. He was moved to paraphrase Marx's quip in "The Eighteenth Brumaire": "Wheat makes its own history, but it does not make it as it pleases."[44] Marx, of course, said this about "men" and not "wheat," but the epigram goes to the heart of the question of historical agency—both human and nonhuman—that Chakrabarty argued was so profoundly at stake in the Anthropocene.

Of course, the Anthropocene is something whose material coordinates are vastly complicated, and go well beyond the relatively straightforward case that I have presented, which is the emergence of temperate cropping zones of global agricultural production. But perhaps, even so, the case is instructive inasmuch as it shows how a commodity (wheat) works its logic into the landscape relentlessly, drawing human actors (settlers) into its incessant production while dispelling other human actors (Indigenous peoples) from the lives they have lived for millennia.[45] As was noted earlier, the new wheat varieties—Red Fife and Marquis (Canada) and Federation (Australia)—made possible the expansion of cropping into areas that were either too cold or too arid to grow wheat that was adapted for European conditions. In doing so, this same commodity—wheat—whose sacramental quality seems to put it beyond question (that is, into the true space of ideology), systematically eradicates ecosystems that have often existed in fairly stable relationships with human communities before the advent of this commodity. On the other hand, perhaps locating agency in the crops that humans have concocted through selective breeding is to fall into a false determinism. Certainly, a Marxist would not be content with ceding historical agency to a commodity, and would rather see the commodification of goods as itself the expression of capitalism and the *class* whose interests are served by its exploitation. It is not wheat, after all, that has privatized the means of production, co-opted the surplus value that is generated by the devaluation of labor, arrogated the accumulated nutrients of the soil, or externalized its ecological costs (soil erosion, insect extinction, soil salinization, nitration of waterways, and so on). In a similar vein, Slavoj Žižek has been critical of Chakrabarty's stance in "The Climate of History," suggesting he was too quick to abandon his Marxist bearings in the search for a species history.[46]

None of the books that I have discussed (*Wolf Willow*, *Wild Card*, *Twice Trodden Ground*) have an especially developed sense of the material determinants of the farms that shaped their lives and that form the traumatic centers of their experience. Each remains partially bound to the exceptionalism that sustains the aura of their authors' childhood worlds even as they subject these worlds to painful and exacting critiques. But what they do offer is a model for the kind of history that Chakrabarty suggests is necessary for the Anthropocene. What they exemplify is not a new synthesis—no new "universal" actor or dialectical protagonist that might form the basis of a *species history*—but a prototype for what it looks like to write from a position of radical antagonism. By mobilizing the constitutive split of subjectivity, which each of them dramatizes as an impossible encounter between a rueful adulthood and the site of their origins, Stegner, Hewett, and York Main provide a mechanism for coping with the epistemological consequences of the traumatic collapse of natural history into human history. Of course, on its own, this does not solve anything. The ecological problems we are causing will continue apace whether we choose to look at them or not. But at least, by following this method, we might begin to be able to confront catastrophe from a place within ourselves where collapse is a condition and not a specter. It is this point of collapse (of *collapsing*) which our three memoirists have drawn with such consistent precision and which provides a paradoxical grounding for "a universal that arises from a shared sense of a catastrophe" and the subjective basis for anthropogenic restitution.

The University of Western Australia

Notes

1. Morton, reflecting on the incessant data of climate change and ecological crisis, puts it this way: "Right now it's as if we are waiting for just the right kind of data, then we can start living in accord with it. But this data will never arrive, because its delivery mode is designed to prevent the appropriate reaction—we find ourselves in the midst of horribly confusing, traumatic events such as global warming and mass extinction, and we don't have much an idea of how to live that." Morton, *Being Ecological*, 15. In a similar vein, Latour writes, "It doesn't stop; every morning it begins again. One day, it's rising water levels; the next, it's soil erosion; by evening, it's glaciers melting faster and faster; on the 8 p.m. news, between two reports on war crimes, we learn that thousands of species are about to disappear before they have even been properly identified." Latour, *Facing Gaia*, 7.
2. Lynch, "Eco-Memoir"; White, "Eco-Memoir." Parts of this essay draw on the paper "Wolf Willow and Wild Card" that I delivered at the 2017 Western Literature Association in Minneapolis. Short sections of this essay also derive directly from passages of my book *Like Nothing on This Earth: A Literary History of the Wheatbelt* and an earlier essay, "Islands of Yesterday: The Ecological Writing of Barbara York Main."
3. Chakrabarty, "The Climate of History," 206.
4. Chakrabarty, "The Climate of History," 197.
5. Chakrabarty, "The Climate of History," 208.
6. Chakrabarty, "The Climate of History," 197.
7. Chakrabarty, "The Climate of History," 218.
8. Chakrabarty, "The Climate of History," 219–220.
9. Chakrabarty, "The Climate of History," 220; emphasis added.
10. Chakrabarty, "The Climate of History," 221; emphasis added.
11. Couser, *Memoir*, 21.
12. Driscoll and Gilmore, *Witnessing Girlhood.*
13. Chakrabarty, "The Climate of History."
14. Huff, "After Auto, after Bio," 280.
15. Fedak, "Marquis Wheat."
16. Wrigley, "Farrer, William James."
17. Belich, *Replenishing the Earth.*
18. Indeed, York Main does not recall meeting a single Aboriginal person in her years growing up near Tammin.
19. York Main, *Twice Trodden Ground*, 1.
20. York Main, *Twice Trodden Ground*, 2.
21. York Main, *Twice Trodden Ground*, 1.
22. York Main, *Twice Trodden Ground*, 2.
23. Chakrabarty, "The Climate of History," 222.
24. Chakrabarty, "The Climate of History," 222.

25. Hewett, *Wild Card*, 11.
26. In reading Hewett's memoir now, one must note that there are traumatic events that in turn inflect the home which she created with Merv Lilley and their daughters Rozanna and Kate. In 2018, revelations regarding Hewett's and Lilley's awareness, and even approval, of the sexual abuse of their teenage daughters were made public. Rozanna Lilley's *Do Oysters Get Bored?*, for example, explores the generation—and generations—of abuse on both their home and the nation more generally. Although *Wild Card* does not cover the years of her daughters' abuse, it was written in their wake, and the Lilley daughters' writing, emboldened by the #MeToo movement (Jervis-Read, "'With Complexity'"), speaks back to the self-fashioning of their mother, creating a series of life writings that track trauma. In the wake of these revelations, Hewett's determination to create a mythological "self" in her writing—for example, this ten-year-old child, precociously aware of the sexual desire she evokes—is center stage in readings of *Wild Card*. See Sutherland, "In Rozanna Lilley's Memoir."
27. Stegner, *Wolf Willow*, 19.
28. Stegner, *Wolf Willow*, xvii.
29. Hewett, *Wild Card*, 11.
30. Thacker, *The Great Prairie*, 190.
31. Stegner, *Wolf Willow*, 8.
32. Stegner, *Wolf Willow*, 9.
33. See Albrecht, "'Solastalgia.'"
34. Chakrabarty, "The Climate of History," 218.
35. Being able to identify (empathize) with someone or something is often now treated as an independent faculty that one either possesses by endowment or acquires through cultivation in much the same way that "sensibility" was thought to work for the romantics.
36. Lacan, "The Mirror Stage as Formative," 1–8.
37. For a discussion of the different "I"'s of autobiography, see Smith and Watson, *Reading Autobiography*.
38. Hewett, *Wild Card*, 271.
39. York Main, *Twice Trodden Ground*, 2.
40. York Main, *Twice Trodden Ground*, 37.
41. York Main, *Twice Trodden Ground*, 42.
42. The idea of the narrating "I" is discussed in Smith and Watson.
43. York Main, *Twice Trodden Ground*, 42.
44. "Men make their own history, but they do not make it just as they please; they do not make it under circumstances chosen by themselves, but under circumstances directly found, given and transmitted from the past." Marx, "The Eighteenth Brumaire," 594.
45. A brilliant example of a history which tracks the structuring effect of an agricultural commodity is Beckert's *Empire of Cotton*.
46. Žižek's critique of Chakrabarty was made in his book *Living in the End Times*, 332–334. Chakrabarty has since responded in "Climate and Capital."

Disclosure Statement

No potential conflict of interest was reported by the author(s).

Works Cited

Albrecht, Glenn. "'Solastalgia': A New Concept in Health and Identity." *PAN: Philosophy Activism Nature* 3 (2005): 41–55.

Beckert, Sven. *Empire of Cotton: A New History of Global Capitalism*. New York, NY: Penguin, 2014.

Belich, James. *Replenishing the Earth: The Settler Revolution and the Rise of the Anglo-World, 1783–1939*. Oxford: Oxford UP, 2009.

Chakrabarty, Dipesh. "Climate and Capital: On Conjoined Histories." *Critical Inquiry* 41, no. 1 (2014): 1–23.

—— "The Climate of History: Four Theses." *Critical Inquiry* 35, no. 2 (2009): 197–222.

Couser, G. Thomas. *Memoir: An Introduction*. Oxford: Oxford UP, 2012.

Driscoll, Beth, and Leigh Gilmore. *Witnessing Girlhood*. New York, NY: Fordham UP, 2019.

Fedak, George. "Marquis Wheat." *The Canadian Encyclopedia*, April 23, 2013. https://www.thecanadianencyclopedia.ca/en/article/marquis-wheat

Hewett, Dorothy. *Wild Card: An Autobiography, 1923–1958*. Ringwood, VIC: McPhee Gribble, 1990.

Huff, Cynthia. "After Auto, after Bio: Posthumanism and Life Writing." *a/b: Auto/Biography Studies* 32, no. 2 (2017): 279–282.

Hughes-D'Aeth, Tony, Delys Bird, and Dennis Haskell, eds. "Islands of Yesterday: The Ecological Writing of Barbara York Main." *Westerly* 53 (2008): 12–26.

Hughes-d'Aeth, Tony. *Like Nothing on this Earth: A Literary History of the Wheatbelt*. Nedlands, WA: U Western Australia P, 2017.

——. "Wolf Willow and Wild Card." Paper presented at the Western Literature Association Conference, Minneapolis, MN, October 25–28, 2017.

Jervis-Read, Jane. "'With Complexity': Growing up with Dorothy Hewett." *Meanjin*, August 20, 2018. https://meanjin.com.au/blog/with-complexity-responding-to-kate-and-rozannas-stories-growing-up-with-dorothy-hewett/

Lacan, Jacques. "The Mirror Stage as Formative of the I as Revealed in Psychoanalytic Experience." In *Ecrits: A Selection*, Translated by Alan Sheridan, 1–8. Routledge, 1999.

Latour, Bruno. *Facing Gaia: Eight Lectures on the New Climatic Regime*. Translated and edited by Catherine Porter. Cambridge: Polity, 2017.

Least Heat-Moon, William. *PrairyErth: A Deep Map*. Boston, MA: Houghton Mifflin, 1991.

Lilley, Rozanna. *Do Oysters Get Bored? A Curious Life*. Perth, WA: U of Western Australia P, 2018.

Lynch, Tom. "Eco-Memoir, Belonging, and the Settler-Colonial Poetics of Place Identity." *Paper presented at the International Conference on Ecopoetics*, University of Perpignan Via Domitia, July 22–25, 2016. *Tomlynch.net*, July 22, 2016. https://tom-lynch.net/braided-channels-blog/

Maher, Susan Naramore. *Deep Map Country: Literary Cartography of the Great Plains*. Lincoln, NE: U of Nebraska P, 2014.

Marx, Karl. "The Eighteenth Brumaire of Louis Bonaparte." In *The Marx-Engels Reader*, 2nd ed., edited by Robert C. Tucker, 594–617. New York, NY: Norton, 1978.

Morton, Timothy. *Being Ecological*. New York, NY: Penguin, 2018.

Smith, Sidonie, and Julia Watson. *Reading Autobiography*. 2nd ed. Minneapolis, MN: U of Minnesota P, 2010.

Stegner, Wallace. *Wolf Willow: A History, a Story, and a Memory of the Last Plains Frontier*. New York, NY: Penguin, 1990.

Sutherland, Katie. "In Rozanna Lilley's Memoir, a Curious Life Gets Even More Curiouser." *The Conversation*, July 3, 2018. https://theconversation.com/in-rozanna-lilleys-memoir-a-curious-life-gets-even-more-curious-99137

Thacker, Robert. *The Great Prairie: Fact and Literary Imagination*. Albuquerque, NM: U of New Mexico P, 1989.

White, Jessica. "Eco-Memoir: Protecting, Restoring, and Repairing Memory and Environment." In *Mediating Memory: Tracing the Limits of Memoir*, edited by Bunty Avieson, Fiona Giles, and Sue Josephy, 141–156. New York, NY: Routledge, 2018.

Wilson, Jerry. *Waiting for Coyote's Call: An Eco-Memoir from the Missouri River Bluff*. Pierre, SD: South Dakota State Historical Society, 2008.

Wrigley, C. W. "Farrer, William James (1845–1906)." *Australian Dictionary of Biography*. Accessed November 7, 2019. http://adb.anu.edu.au/biography/farrer-william-james-6145

York Main, Barbara. *Between Wodjil and Tor*. Milton, QLD: Jacaranda, 1967.

—— *Twice Trodden Ground*. Milton, QLD: Jacaranda, 1971.

Žižek, Slavoj. *Living in the End Times*. New York, NY: Verso, 2010.

"As Closely Bonded as We are:" Animalographies, Kinship, and Conflict in Ceridwen Dovey's *Only the Animals* and Eva Hornung's *Dog Boy*

By Grace Moore

ABSTRACT

Using the fiction of Ceridwen Dovey and Eva Hornung, this essay considers animalography as a medium to represent animal emotions, particularly when ties of kinship break down. It addresses the difficulties and power dynamics associated with speaking for nonhuman others, while engaging with Cynthia Huff's cautions regarding the posthumanist life narrative.

> So what kills animals? Everyday human activity. The stuff that keeps us alive and the stuff that makes those lives worth living. The stuff we do even though it makes none of us happy (war). Living in houses and eating food and going to work and wearing clothes and making movies and books and wine and theatre.
>
> — Jane Rawson, "The Invisible Extinctions"

Animal and human lives have always been inextricably intertwined. With some species, these connections are apparently ties of companionship, although usually not on equal terms. For other animals, the impact of human dominance has been catastrophic: countless species are losing or have lost their habitats as a result of human "progress," while others are born into captivity, their breed having been identified as "livestock" generations ago. Even in the most comfortable companionate situations, the relationship between human and animal is fundamentally unequal, with the human making non-consultative decisions about the animal's diet, reproductive capacity, and contact with others of their kind. The animal rights organization PETA (People for the Ethical Treatment of Animals) has compared keeping a pet to keeping a slave, and while some would dismiss their rhetoric as sensationalist, the relationship is a vexed one.

The power dynamics of the human|animal relationship are frequently skewed to the detriment of the nonhuman animal, while even in the most positive scenarios, interactions tend to be inflected by anthropomorphism. We see *ourselves* and our own emotions in the animals around us, and while we may believe that we understand their desires and needs, this belief is all too often entangled with ideas of ownership, or a kinship that is not grounded in equality. Taking two very different literary examples, Ceridwen Dovey's extraordinary short-story collection *Only the Animals* and Eva Hornung's confronting and original novel *Dog Boy*, this essay will consider how fiction and biography can advocate for animals. It will engage with the role animalography can play in illuminating cross-species narratology, while also addressing some of the shortcomings associated with speaking on behalf of an animal. In so doing, it will take on board Cynthia Huff's reminder of the challenges involved in "realizing that transhuman is not posthuman" and the need to "consider the representational shortcomings of popular posthumanist life narrative."[1] Furthermore, it will consider the limits of our empathy with the nonhuman other, while at the same time attending to the permeability of the boundaries that Western society has erected between the animal and the human, particularly during times of food scarcity.

Dovey stages the difficulty of writing the life story of a nonhuman in her collection of animal autobiographies, *Only the Animals*. Each of her ten stories features an animal recounting the circumstances of her or his death (all of these deaths come at the hands of humans). Dovey's definition of the "animal" is broad insofar as one of her tales features a mussel as a protagonist, challenging preconceptions that her readers may hold about the sentience of particular beings, along with hierarchical beliefs in the superiority of one species over another. In so doing, she engages with Jeremy Bentham's oft-repeated assertion "The question is not, *Can they reason*?, nor *Can they talk?* but, *Can they suffer*?" while also demonstrating that neglect and cruelty to animals can take a range of different forms.[2] Dovey's stories explore the emotional lives of her chosen creatures, focusing particularly on the issue of betrayal. Hornung takes a very different approach in *Dog Boy*, inverting our usual expectations of the relations between dogs and people through her powerful depiction of a young boy who joins a canine pack and gradually assimilates himself into a nonhuman existence. Both Dovey and Hornung consider what Barbara Creed reminds us is the "shared history and fate of human and nonhuman animals," subtly undercutting the anthropocentric nature of that history by asserting the importance of the animal perspective and refusing to privilege the human.[3]

It is, of course, paradoxical to conceive of an animal autobiography (or "animalography," as Cynthia Huff and Joel Haefner have termed it[4]) that is written by a person, and the séance-like quality of Dovey's writing reveals her consciousness of this challenge. Animalographies attempt to engage with this question by translating animal emotions into recognizably human terms, but, as I shall discuss in this essay, they constantly do battle with allegations of anthropomorphism and ventriloquism. Donna Haraway helpfully argues for a compulsion that transcends narratology when she writes of the "flesh-to-flesh and face-to-face" inheritance of histories through touch, which drives her to recount the "breed stories" of the Australian Shepherd dogs with whom she shares her life.[5] As Haraway demonstrates, in spite of its limitations, animalography continues to play an important role in fostering interspecies empathy.

Dovey highlights what Freya Mathews has helpfully termed the "estrangement" of humans from the animal world,[6] while at the same time compounding that remove by having her animals tell their stories from the afterlife. Each creature is lucid and matter-of-fact, with their subjectivity apparently unchallenged by moving from one life to the next. These animals express themselves in language now that they are dead, although in their lifetimes their apparent inability to communicate with people (or, it is suggested, human obtuseness) has rendered them as inscrutable to their human observers as Jacques Derrida's cat.[7] This inscrutability raises difficulties when it comes to telling stories *of* and *for* animals. As Huff has noted, "foregrounding the subjectivities of nonhuman animals and machines by granting both groups human rights (including the right to tell one's own story, to have a voice) has been problematic precisely because neither nonhuman animals nor machines possess a human voice or experience their worlds via the technology of natural human language."[8]

Dovey purports to allow each of her creatures to tell their own tale, but she emphasizes the inevitably anthropocentric nature of her task by aligning each animal with a human, whose presence gradually impinges itself on the account. As I discuss later in this essay, Hornung similarly engages with the complexities of interspecies communication, partly through recognizing that speech has been privileged over other forms of sensory interaction, such as smell, and partly through an interrogation of the propensity to project emotions onto animals.

Dovey's collection is framed by a concern with human-engineered conflict, although it looks beyond the conventional emphasis on casualty figures that apply to people to consider the effects of war on other members of the nonhuman world.[9] In so doing, Dovey subtly undermines human pretensions to "mastery," but she also reminds readers that, for many species, daily life can be akin to an ongoing battle, or rather an onslaught waged by an adversary with technology on its side and a strong sense of entitlement. Claire Hielscher encapsulated the collection's emphasis on the wide-scale suffering caused by humanity in an early review when she remarked, "Each story takes place at the nexus of a destructive human conflict some time between 1892 and 2006: World War I, Pearl Harbor, Mozambique's civil war, the conflict in Iraq and many others. The creatures' responses to the conflicts they are caught in add an empathetic layer to the battles: their pity of humans, their confusion, jealousy, heartache and sense of impending loss imbue momentous events with an even heavier sense of gravity, forcing readers to assess the actions and impact of the human species."[10] Yet I would argue that each animal's story amounts to more than an "empathetic layer" that accentuates human emotion; rather, they collectively draw attention to the self-centeredness of humans, whose investment in interspecies companionship breaks down as soon as conventional food sources run out. Like Hornung, Dovey troubles the idea of cross-species kinship. Both authors show how conflict can raise questions about which lives are valuable and deserving of preservation.

In Dovey's work, the animal responses to the treachery of those whom they have loved and trusted are often devastating in their intensity and bewilderment. Each of her stories showcases the loyalty of different animal species toward humans who are less stalwart in their affections. Almost all of the animals identify themselves in relation to a well-known literary figure. The presence of notable literati, including Franz Kafka and Virginia Woolf, purposely skews the collection's titular claim to be "only the animals," drawing in the human, possibly to make the featured creatures additionally remarkable or possibly to show how even a greatly admired cultural figure can willfully misunderstand animal alterity. Dovey may be asserting the animal qualities of the human animals behind the stories or she might be attempting to demonstrate how even the most sensitive of human voices inevitably impinges on the animalography.

The question that lies behind the title of Dovey's collection is elucidated by the epigraph to "Telling Fairytales: Soul of Bear," the eighth story, which expands on a quote from the American academic Boria Sax, who has worked extensively on human representations of animals: "What does it mean to be human? Perhaps only the animals can know."[11] Dovey's "knowing" animals thus shed light on the inequity of human-animal relations, demonstrating that the anthropocentric version of events is not necessarily an accurate one. As Dovey makes clear, the assorted animals who tell their stories have important and deeply affecting counter-narratives to put forward, which undermine human claims to dominance and frequently highlight cruelty and carelessness on the part of "owners" or "trainers." Dovey thus attempts to place animal emotions at the heart of her work, yet she also registers the difficulties and contradictions that inevitably accompany such a task.

A question that underpins a consideration of the enigmatic animal must be why it is important to attempt to capture their apparently elusive emotional states. Erica Fudge has gone some way to answer this concern with her remark that "[s]tories might allow us, as readers and thinkers, to find a path between the impossible and the unnecessary. The imagination might offer a way of bridging the chasm that seems to separate humans from animals."[12] While Fudge is right to consider the importance of story-telling in generating empathy for the—at times—unfathomable beast, these animal stories are another version of speaking *for* the animals and inevitably obscure the actual tales their subjects might tell if humans knew how to listen. This problem is apparently insoluble. Even those animals with whom we live in the closest proximity remain emotionally mysterious, and while we may project inner lives onto them, we cannot yet truly understand their thoughts and feelings. David Herman encapsulates the ethical bind that surrounds the imposition of a human voice onto a nonhuman "subject" when he remarks that, "whereas attributing voices to nonhuman animals can be viewed as a form of 'connection and inclusion,' unless it is situated within what Huff and Haefner would describe as a critical-posthumanist versus popular-posthumanist perspective, this narrative practice raises the specter of paternalism. Such voicing also poses the threat of a subordination of difference under the logic of the same—an emptying out of the specificities of experience that can obscure relational ties across species lines—cautioned against by postcolonial theorists and feminists in intraspecies contexts."[13] Dovey's stories consider this dilemma, which she signals through the use of provocative subtitles such as "Dumb Animals" and "Neighbours."[14] Moreover, her most overt engagement with the issue of animalography is in her story of a dolphin, a creature celebrated for its empathetic communication with humans.

The dolphin, who dies in military service in Iraq, pens a posthumous letter to the poet Sylvia Plath. The missive begins with a critique of Plath's former husband Ted Hughes' celebrated animal poetry: "I realised, as I read them [the poems] again, that I had misunderstood them on first encounter. Back then, I had admiringly thought he was trying to understand the human by way of the animal, but now I can see in fact he wanted to justify the animal in the human. I saw right through his mythologising of the poetic process, the animal as symbol of the poet getting in touch with his deepest, wildest, most predatory instincts."[15] The dolphin, known as "Sprout," registers the uses to which animals are put and notes that even the most powerful evocations of the animal other are, in fact, relentlessly self-serving. Hughes, she implies, drew on the animal to justify his own impulses, and she identifies a "relentless maleness" to his voice and also, by inference, to his behavior.[16] Lest we miss the paradox that Dovey exposes in this scene, where she too projects human emotions onto a nonhuman protagonist, Hughes is defended by Sprout's friend, the "author and philosopher of sorts" Elizabeth Costello.[17] Costello's riposte is, of course, in part a winkingly metafictional acknowledgment of the work of *her* creator, the novelist J. M. Coetzee (who, like Dovey, is a South African-Australian), and the inevitable parallels to be drawn between the two writers and their shared interest in animal sentience.[18] Yet the imaginary Costello's comments also suggest that while penning an animalography may seem futile or self-indulgent, in terms of a cross-species understanding, it might be better to write misguidedly from an animal perspective than not to make the attempt at all: "It is an attitude that's easy to criticize, to mock, … It is deeply masculine, masculinist. Its ramifications in politics are to be mistrusted. But when all is said and done, there remains something attractive about it at an ethical level."[19] This comment hangs over the remainder of the story, which is remarkable not only for its representation of the connections between Sprout and her human "trainer," but also for its clear assertion of the dolphin's intelligence and empathy, which is far greater than that of the story's human characters.

The evolutionary biologist Marc Bekoff stresses some of the restrictions that humanity faces in speaking on behalf of animals with his comment that, "Our anthropocentric view of other animals, in which humans are so taken with themselves, is far too narrow. The worlds and lives of other animals are not identical to those of humans and may vary from species to species and even within species. The same problems arise in the study of emotions … if we believe that emotions in animals are going to be identical to or even recognizably similar [to those] among different species."[20] Bekoff points out that a further barrier to understanding animal emotions is that we frequently encounter animals in environments which have been altered by humans, making it difficult for them to reveal what he terms "their own form of genuine morality."[21] Yet, like Dovey, he registers that there is a value to the anthropomorphic, even while it poses an obstacle to truly knowing the animal other. As he remarks, "By engaging in anthropomorphism we make the world of other animals accessible to ourselves and to other human beings. By being anthropomorphic we can more readily understand and explain the emotions or feelings of other animals."[22] Herman ventures further, in an analysis of Dovey's "Letter to Sylvia Plath," to posit the idea of "co-authored acts of narrating," which he regards as enabling "solidarity-building projections of other creatures' ways of being-in-the-world—projections that enable a reassessment, in turn, of forms of human being."[23] Enticing though Herman's argument may seem, its essential difficulty is that he makes claims for a reciprocity that cannot be proven and which valorizes the human scribe's capacity to capture animal emotions in an authentic and meaningful way. For Herman, the moments in the text when Dovey's animal narrator emphasizes a particular gesture or practice—such as when the dolphin speaks of echolocation—allow the animals to speak for themselves. My difficulty with this position is its fixation on the apparent meaningfulness of isolated motions or apparent signals (as is the case when Herman writes of Dovey's depiction of elephant grief rituals) and their interpretation from an anthropocentric perspective. No human depiction will ever capture the subtleties of nonhuman intra-species interactions, and to suggest that depictions of them are tantamount to joint authorship is disingenuous. Such representations are inevitably selective and can, by their very nature, focus only on what appears significant to the human observer and focalizer. This is not to suggest that such representations lack value, as Val Plumwood expresses it: "If anthropomorphism means writing an active, intentional, ecological world, and expanding our range of sensitivity to nonhuman actors and elements, it would be highly rational, especially in the current dire ecological situation, to have more of it."[24] Moreover, as Huff has observed, "[f]or a posthumanist approach to life-narrative scholarship to take hold, scholars must counter ventriloquized representationality in favor of a politics of subjectivity that emphasizes relationality and process among disparate types of beings."[25] Dovey's stories explore exactly this inequality of representation, examining the stories' one-sidedness and emphasizing the confusion that each animal feels when contracts of kinship are breached. Often, as in the story of Colette's cat, who meets her death in the trenches of France during World War I, Dovey's animals attempt to excuse human pragmatism, suggesting that their abandonment is the result of a mistake or some form of misunderstanding. Thus, nonhuman loyalty is pitted against a more expedient attitude to kinship and the belief that ties between species can be severed at times of duress.

The boundaries between the human and animal worlds are, in spite of human suggestions to the contrary, remarkably thin. Dovey and Hornung both explore the permeability of these borders. *Only the Animals* addresses emotional attachments, while *Dog Boy* considers the degree to which species are separate through its depiction of a human boy who is able to *become* a dog. Human stories about animals all too frequently emphasize animal subordination—a convention that both Dovey and Hornung attempt to subvert through their use of the animalography. Fudge has argued that "it is impossible to understand human history without also understanding the role that animals played in that history because animals have always held key positions in so-called human society."[26] She goes on to note that "[i]n the history of animals, humans are regarded as inseparable from animals and not as a species in isolation. Thus, paradoxically it might seem, this more complete history of humanity is also a history that undoes the idea of the separate and superior status of the human and this in turn changes the nature of history itself."[27] While Fudge is absolutely right to emphasize the inseparability of people and animals throughout history, her assertions of coexistence sit uncomfortably alongside notions of the "utility" of the animal. Feminist ecocritics, including Plumwood, have highlighted the discourses of "mastery" surrounding the natural world, which license ideas of dominion over nonhuman beings and their treatment as "resources."[28] This sense of a divinely mandated hierarchy, with humanity at its head, makes it easy to consider animals and plants as expendable, or as property to be "managed" and farmed. What should, perhaps, be surprising within this worldview is that *any* animals are considered to be friends, rather than assets. However, the notion of "charismatic nature" goes some way to explain the selection of animals who inhabit the home *versus* those who are treated as resources. Both Dovey and Hornung consider what it means for animals to be both domesticated and "wild," and they interrogate the associations of these terms.

While each of Dovey's stories explores human responsibility for animal death, a number of them directly confront the vexed question of who is kin and who is food. Dovey examines the willed destruction of the boundary between companion species and livestock, giving each animal a voice through which to tell her or his story. The animals' power as storytellers allows a number of them to scrutinize the moment at which their companion status is compromised and lost. Several of the stories feature poignant moments at which animals realize their betrayal, although perhaps one of the most chilling is Dovey's reworking of Kafka's "A Report to an Academy."

In "Red Peter's Little Lady," the primate from Kafka's original story, Red Peter, rejects the companionship of Hazel, a chimpanzee who has been chosen as his bride (probably the "small half-trained chimpanzee" with whom Red Peter "takes his pleasure" in Kafka's short story), in favor of Frau Evelyn Oberndorff, a married woman who has taken over Hazel's care.[29] The story comprises correspondence between Evelyn and Red Peter, with occasional interventions that purport to be from Hazel but which are transcribed by Evelyn and reveal the female ape's discomfort at attempts to render her more human. Initially, Peter's affection seems to be reciprocated by Evelyn, whose letters include suggestive (and slightly absurd) phrases such as, "And yes, if you are wondering—I do remember that night, reading Buber together, and everything else," which imply that the pair have allowed the boundaries between species to relax in ways that are about more than mere sociability.[30] However, as food supplies dwindle and rationing becomes more and more severe, Evelyn's letters become increasingly focused on the practicalities of feeding her children. Peter may long to see her and, it is implied, to return to enjoying carnal pleasures with her, but Evelyn's missives obsessively focus on food:

> Dear Red Peter
>
> I do not want to worry you—we are fine, mostly. But food is scarce, as you say, and we are struggling to find enough to feed the animals at the zoo. The children and I are surviving on dark break, a few slices of sausage with no fat, and three pounds of turnips a week.[31]

As the weeks go by, Peter begs Evelyn for her company, while Hazel critiques the human reliance on words to communicate, imagining Peter tossing words to her when, as she puts it, "I swing from the curtains towards you and display my arsehole"—a much more direct and effective form of communication than the equivocating letters.[32] Hazel continues to ask a question that resonates through the story's closing pages when she demands of Peter, "Would you like me to be more human, or less human, or more or less human?"[33]

Red Peter equates being human with refinement and sensibility, yet Evelyn's letters show that she and her family are rapidly degenerating. Initially, Red Peter sends them food, which they devour voraciously and at times without restraint. Yet he is soon unable to procure goods on the black market, noting that "[n]obody wants to see an ape eat when there are humans going hungry."[34] Hazel responds to the scarcity by retreating into her cage and turning herself into a "hunger artist," in a nod to another of Kafka's works. Her gesture is, however, also an act of intelligent resistance, which she hopes will allow her to starve to death rather than being eaten by human predators. Evelyn reports that there are no longer any visitors to the zoo, so there will be nobody to watch Hazel's macabre show. Meanwhile, Evelyn lures Red Peter to visit her, telling him that she can hardly wait to see him. Evelyn's final letter gives way to an anonymous piece of musing, describing the pervasive hunger and penury: "The creatures around me are no longer being fed. They would do anything—anything—for food; in fact they debase themselves daily begging for it. Does it matter now? That some of us will die of disease, some of malnutrition, some of exposure? We will all die hungry, but only I will have *chosen* to starve. The humans are too fragile, held together by not much more than shared food on a table. Is that all that lies between the behaviour of apes and humans? A regular supply of hot meals? … The light turns blue before the window. I hope they heat me up and eat me."[35] This voice may be that of Hazel, since the passage contains a number of references to the "hunger artist" and how his spectacle became commonplace, but it is also possible that the voice is that of Red Peter himself. In his last letter, which comes immediately after this unsettling section, Peter writes to Evelyn with characteristic affection, even after it has become clear that she means to eat him. He reflects on a spoon of pumpkin marmalade that she has offered to him, recognizing it as part of a plan to fatten him up. Red Peter's story ends with the distraught declaration, "immediately I knew I had made a terrible mistake getting back into this cage. Don't let us miss our chance, darling! Unlock this cage, let me out, let me into your bed."[36]

That the letters end so abruptly indicates that Red Peter's fears were just, but his final words show a pathos-filled faith in human nature. He believes that Evelyn can be swayed, just as he attributes this "terrible mistake" to his decision to reenter the cage. As Clare Archer-Lean astutely expresses it, "[a]nthropomorphism becomes tragic in this story."[37] For Peter, it is the act of returning to captivity that renders him an ape once more, yet, for Evelyn, his transition back to the status of an animal occurs long before this symbolic moment. Her exploitation of his generosity, coupled with her conniving to bring him back to her home, reveals a ruthless streak that the lovelorn ape is too compassionate to understand. He has become more human than the humans around him, holding himself to a moral ideal long abandoned by the people of Berlin. Indeed, when it becomes expedient to deny his companion status, Evelyn appears to be unhesitating in her rapid reinstatement of a species boundary between the two of them. Dovey here challenges her readers to think through the ethics of this situation and to register its horror, recalling Haraway's important question about how food security and "multispecies coflourishing" can coexist.[38] Dovey depicts the pragmatism of a woman who justifies the ingestion of a one-time lover by choosing to reinvoke a difference she has previously been complicit in effacing. This story is, because of the physical similarities between primates and humans, one of the more arresting of the collection.

Dovey's exploration of the scope of human treachery is not restricted to her representation of the abject ape. In "I, the Elephant, Wrote This," the story is recounted by the soul of an elephant who died in 1987 in the Mozambican civil war, although it also presents the history of Castor and Pollux, the elephants who were shot and killed as food during the siege of Paris in 1870. The elephant narrator (who is a twin) has a strong sense of ancestry and lineage, and, as a consequence, a deep distrust of human beings. Her daughter and nephew are fascinated by the splendid glamor of the Parisian elephants, and they find it difficult to understand the mother's anger when she finds them pretending to be Castor and Pollux, taking imaginary children for rides around a make-believe zoo.

The elephant narrator stresses the genetic proximity of her species to that of the human, while at the same time drawing understated parallels between her life on the drought-stricken plains and the conditions in Paris more than a hundred years before.[39] As the river dries up in war-torn Mozambique, the older elephants forfeit their claim on the water to sustain their young. We learn that the "adults drank less and less so that the young could have their fill"[40]—a statement that resonates a few lines later when the elephant's sister tells the children, "A zoo … is a very dangerous place for an animal in wartime, for it can mean the difference between life and death for the human inhabitants of a city. But it was not the poor who ate the zoo animals in Paris."[41] She continues to detail the list of animals consumed during the siege (the poor ate bread and died in their thousands), which sits in stark contrast to the altruism of the elephants, who put their young first. The grim inventory also reveals once again the concessions that humans are willing to make to their sense of what may be eaten at a time they perceive to be one of crisis: "She told them that the rich Parisians had started first on those zoo animals they could in good conscience eat, the ones that were not so far from the herbivores usually adorning their porcelain plates: two zebras, the yaks, five camels, a herd of antelope. Then they ate the flamingos and the single adored kangaroo. Next they shot and ate the lions and tigers. The zookeeper, desperate to save his beloved hippopotamus, said he would only sell it for food if someone paid eighty thousand francs … Not even the rich could justify such an expense when there were other zoo animals still to be eaten. They ate their way through the wolf pack, drizzled with deer sauce. Then they ate the passenger pigeons that had so faithfully been transporting secret messages from the French command in Tours."[42] This voracious demand for meat from almost any source demonstrates the arbitrariness of the lines that delineate which animals are "livestock," which are "wild," and which are "companions." The story shows loyalty to be meaningless during times of perceived scarcity and it eerily brings to mind Haraway's caution that "[c]ompanion species cannot afford evolutionary, personal, or historical amnesia."[43] Dovey's elephant narrator captures the abjection of the captive elephants, while at the same time alluding to other creatures whose stories appear in *Only the Animals*. The pigeon's position as a faithful ally is, for instance, reconfigured when it seems that it might be more useful as a source of food. Dovey pushes this quandary to its limits when the narrator, who is more distrustful of people than her sister, details how members of high society overcame their misgivings surrounding feasting on primates. As with the example of the pigeon, which speaks to an earlier story in the collection, "Pigeons, a Pony, the Tomcat and I: Soul of a Cat," Dovey draws connections between both stories and animals, asserting an alternative, dismal kinship that is grounded in a shared experience of human pragmatism or, to put it more bluntly, of being perceived as food. As the elephant notes, "I told them that when it came to the monkeys the Parisians paused. They tried for a while to ignore their hunger. Some wrote editorials to be published in the gazettes, which were distributed around France by hot-air balloon (now that the passenger pigeons were all eaten up) … declaring that sometimes it might be better to starve than to eat the meat of a creature that reminded them in some uncomfortable way of themselves, though they could not yet put their finger on exactly *what* or *why* that was. They did not credit us elephants with the same exceptional qualities, and so they turned away from the monkeys."[44] In order to drive home the continuum of betrayal, Dovey emphasizes the devastation caused again more than one hundred years later in Mozambique: "We began to pass the remains of animals killed and eaten by humans, left beside makeshift cooking fires. At first, only the usual: zebra, wildebeest, buffalo. But then we found a pack of wild dog carcasses and stopped to mourn for them. They are as closely bonded as we are, their packs woven together by mutual affection and trust. We spent time breaking branches to cover their bodies out of respect."[45]

This macabre accumulation of remains recalls Derrida's parallels between animal deaths and human genocide,[46] while the carelessness with which the corpses are abandoned reveals a callous disregard for any remaining vestiges of kinship between the human and animal worlds. Fudge helpfully identifies what she terms the "confusion" experienced by carnivores in relation to animals, suggesting that meat eaters' meal choices are acts of negotiation and renegotiation as they seek to apply a logic or a "rule" to justify their food choices: "It is the meat eaters who live with boundary confusion. They must continually make careful and complex decisions about where the boundary between the edible and inedible exists. Humans cannot be eaten under any circumstances, the argument goes, but sheep can. Dogs, while not human, cannot be eaten, not because they do not taste nice, but because they are categorized differently from sheep: they are pets. Rabbits, however, can be pets and meat (although it is unlikely in most cases that one animal exists in both categories simultaneously), … A close look at the underpinnings of meat eating reveals how insecure the categories of difference it proposes are. What is and is not edible is not a natural opposition: if I can eat a sheep, what stops me from eating a dog, a cat, a chimpanzee, a human?"[47]

Fudge goes on to remark, tongue-in-cheek, that "[v]egetarians, in comparison, seem to have an easy life," and her analysis points to the randomness of choices relating to the status of different species.[48] Sheep are identified as food (and, as I write this essay, truckloads of sheep pass by the library window on their way to an abattoir on the other side of the city), yet as cultural critics, including Philip Armstrong, have pointed out, sheep are highly intelligent, with the capacity to recognize human faces and recall human behaviors.[49] To choose to eat a sheep—or any other animal—involves a denial of its intelligence and capacity to tell stories, as well as its ability to live an emotional life.[50] Mathews follows this choice through to its uncomfortable conclusion when she comments, "In drawing animals into our domiciliary space, and raising them within the circle of the human clan, and then slaughtering them for food or other purposes, we in fact violated the taboo against violence towards kin. The moral gravity of this transgression then required that we rationalize our action by denying the moral significance of domestic—and by extension, other—animals, reducing them to the status of objects that may be produced and consumed without the slightest compunction. In other words, to justify the utilization of animals raised, like kin, within the human domain, we invented an ideology of animals as objects, which effectively closed our eyes to their otherwise manifest subjectivity."[51] As Mathews sees it, animals have been deceived into offering their companionship. Instead of mutuality, they have encountered violence and exploitation, both of which have been justified by the erection of imaginary boundaries to uphold a human sense of dominion over the natural world. That is why, in spite of its limitations, the animalography continues to be an important medium, asserting what Haraway labels "attachment sites" and drawing readerly attention to nonhuman sentience and the capacity for suffering.[52]

Eva Hornung explores a world in which borders between species are less rigid in her novel *Dog Boy*, a work that might be a fictionalized biography but which plays with the possibility that it is an animalography. *Dog Boy* tells the story of Romochka, who, at the age of four, finds himself alone in his apartment, abandoned by his mother and an "uncle" who has clearly been a malign presence in the child's life. Starving and alone, Romochka turns to a stray dog, Mamochka (or "mother"), who has recently whelped, and he gradually joins her pack. Finding care from dogs rather than the human beings who are repelled by him, Romochka assimilates into the pack and increasingly self-identifies with the canine rather than the human. Romochka appears to have been "drawn into the multispecies knots" that Haraway identifies as part of what she terms "becoming" with dogs, so completely has he been drawn into the dog world through "touch" and "regard."[53]

The novel is set, for the most part, in the slum areas of contemporary Moscow, which look almost post-apocalyptic. This is a world in which mountains are made from garbage, while the detritus of society is left to scavenge, with humans eating dogs, and dogs occasionally eating humans. There is little evidence of human kinship in this city, where the homeless face danger from the *militsiya*, from the contemptuous "housed," and from each other, and where almost anything living is perceived as food. Indeed, at several points in the novel, when Romochka must make strategic decisions about whether to align himself with particular people or creatures, he repeats a phrase of his "uncle" to remind himself that they are "the only dish on the table," thus couching his social interactions—and survival—in terms of food.[54] Hornung is unflinching in her graphic descriptions of how the stray dogs and the stray boy hunt and feed. Romochka quickly assumes the behavioral characteristics of a dog, self-consciously at first and then later—when he comes to feel a love and acceptance from the dogs that he has never experienced from his human family—as a matter of course. A key moment in which Hornung registers the boy's transformation appears early in the novel, when he and "Black Sister," one of the puppies in his new pack, hunt a rat together. Having previously depicted him ransacking his apartment for food, and balking at the challenge posed by a jar with a tight lid, Hornung now shows a child who has adapted so rapidly that he thinks nothing of stalking and killing vermin.

While, for Romochka, the hunt has become second nature, for the reader who has internalized the social taboos discussed by Fudge and Mathews surrounding which animals may be considered food, the scene evokes a highly visceral response, bordering on nausea. The narrator captures the shared excitement of dog and boy, disregarding accepted sensibilities surrounding appropriate forms of food and eating: "Black Sister was watching him. She reached a polite paw towards the rat, looking bright eyed at him. He handed it over and she pulled the rat open with great delicacy. He was delighted. They lay down on their bellies and spread it out. This was, he decided, his favourite food. He chewed through the slippery ribcage to its soft centre, keeping the head in his fist to make sure Black Sister didn't crunch through it and eat his treasure."[55]

Hornung compounds the power of this scene a few pages later when she shows Romochka drawn to a fire, even though he knows he cannot join the humans gathering around it. We learn that "[h]e liked fires but had never been able to get near them. People here knew that he was not one of them, and he guessed that he broke their rules, crossed their invisible closed paths and offended their order."[56] This tableau is thought-provoking in part because it shows how swiftly Romochka has been exiled from human society, but also because it shows the randomness of his exclusion. At the apparently convivial campfire, the men are roasting a dog: "The music swelled and reeled him in. The men were lit up in orange light, roasting the hindquarters of a slaughtered dog. Its pelt and head lay on the stained snow. It wasn't a dog he recognised, and he thought from the smell that it was probably one of theirs, not a clan dog. Someone would be wearing that golden pelt in a few days."[57] The fact that the dog was "one of theirs" suggests that it is a companion animal who has—through necessity, actual or perceived—crossed the border separating kin and food. Yet Hornung also takes the opportunity to remind readers of Romochka's biological difference from his pack when her narrator comments, almost casually, "The smell of cooking meat was very good."[58]

The campfire scene is a complicated one, partly because Romochka has already begun to assimilate himself into the pack by this point in the narrative. It therefore evokes a repulsion that transcends Western society's usual disdain for cooking and eating dogs, and borders on the cannibalistic. This moment, when Romochka is tempted by the aroma of roasting flesh, is one of several in the novel where the boy is shown to be on the brink of dissolving his newfound kinship with the dogs in order to revert to human-style predation. The porosity of this boundary is underscored by several other fleeting references to cannibalism, including one from the end of the work where Natalya—a pediatrician who is attempting to domesticate Romochka—visits the lair he has shared with the dogs. Among the remnants of the items scavenged by the dogs and boy, Natalya spies a pile of bones: "They looked like big dog skulls. Did they eat each other? That idea pulled another out of the tumbling darkness in her chest. She tried not to look further into the shadows, suddenly fearful that there might be human bones there too."[59] That the idea of dogs (and boys) eating dogs is so rapidly followed by a notion of dogs (and boys) eating humans testifies to Natalya's horror at the consumption of a companion species. It also points to her awareness of how Romochka has transgressed both the borders of kinship and those of dietary taboos. In an earlier scene, the confusion of his allegiances becomes clear when he kills a stray dog who has been stalking him in a harrowing incident that ends with the words, "He had never before killed one of his own kind," before shifting immediately to a sighting of "uncle," which leads him to kill the man in anger.[60] The pack is horrified by this latter action, which is a vicious murder rather than self-defense, and which speaks to the blood-thirsty to avoid repetition human conflict underpinning Dovey's short stories.

There are a number of instances in the novel where Romochka, in spite of his outcast status, is drawn to his own kind, often with consequences that pose problems for the power dynamics of his pack. The humans he encounters are almost universally convinced that he is not one of them. A marginalized woman with a scarred face is startled by his sudden appearance and, through free indirect discourse, the reader learns that "[e]veryone here knew he was not one of them, that he was wild and had dogs."[61] The notion of Romochka "having" or "owning" the dogs recurs throughout the novel, as characters find it impossible to consider the equal terms on which the child lives with them. Indeed, there are moments when Romochka himself forgets his position as subordinate to (and dependent upon) Mamochka, the matriarch of the pack, and attempts to assert dominance or "mastery" over the dogs, suggesting that there are aspects to his conduct that were either acquired before the age of four or which are, more troublingly, an innate aspect of human dominance.

Romochka's association with the dogs is a source of fear rather than wonder to those who meet him on the streets. When he abandons conventional hunting to steal a woman's shopping, she responds with revulsion, shrieking "Filth! Bomzh! Animal!"—collapsing dirt, dispossession, and animalism together in a way that continues to trouble the boy after the encounter.[62] He later dwells on the confrontation, suddenly conscious of himself as an outsider: "'Filth. Bomzh. Animal'. He said the words over and over to himself, his own rusty little-kid voice scaring him in the darkness … Something dissolved suddenly, some huge barrier between him and people. They had been in the untouchable fringe of his world; dangerous, like dogs, with frothy mouths, yet irrelevant like those sick dogs, too. He started to cry softly to himself."[63] This scene is remarkable for its pathos and abjection while also evoking Mary Douglas's oft-quoted assertion about dirt as "matter out of place." As Douglas expresses it, in labeling something as "dirt," "we are using a kind of omnibus compendium which includes all the rejected elements of ordered systems."[64] Romochka is rendered "dirty" because he lives not only *with* dogs but also *as* a dog, although his displacement has involved a degree of choice. The boy who has mutated into a dog has also transformed into a nuisance, and the local people are quick to strip away his kinship. The narrator informs us: "He is a demon, some say, who eats the flesh of humans and wanders alone in the form of a child to tempt people near."[65] Later, we are told, "[h]ouse people … would hate to touch him, would never think to jump him, hold him down or beat him."[66] In theorizing the choices that humans make regarding which animals are kin to be cherished and which are vermin to be destroyed, Deborah Bird Rose has commented, "Many thoughtful persons have noted that once an animal is declared 'pest', or 'vermin', or even 'invasive', something happens within the sensibilities of many humans."[67] Labeled a "pest" from an early age, Romochka embodies the human dimension of this selection process; through being suckled by a dog and spurned by humans, he challenges regular kinship structures and hierarchies.

While the city's residents are repelled by the "feral boy" that Romochka becomes, he identifies himself almost proudly as a "stray." When a kind cook offers him food, he responds to her comment that she thought Mamochka was a stray by declaring, "Yes … Me Too."[68] The concept of the stray is often used as a form of control, however. Early on in the novel, Romochka and his pack elude a government campaign to "eradicate" strays, while later Grey Brother is captured and castrated because of his stray status. Barbara Creed notes how capacious the term can become when it is used to exclude, observing that "[t]he project of erecting boundaries to keep the other (strays, animals, refugees, indigenous peoples, exiles, women, the poor) at bay in order to assert the primacy of the so-called civilized human animal over all others signifies a failure of empathy that now threatens to undermine all forms of life in the Anthropocene."[69] To decree that an animal or a person is a "stray" is, according to Creed, a means of absolving oneself of social responsibility for them. The failure of empathy that she outlines here is analogous to the breakdown of kinship that allows the humans who encounter Romochka to see a "stray" rather than a child. This absence of compassion is clear when the boy is apprehended by the *militsiya*, and one of the soldiers declares, "Feral kids are worse than rabid dogs. Worse than adults too, and they reckon there's millions. Never solve anything unless we get rid of them."[70] He goes on to deny Romochka's humanity, asserting, "That's no kid. That'll kill my kids given half a chance."[71]

Society's indifference to Romochka and the many children like him across the city does not prevent the boy from seeking companionship. He is drawn to his own kind and to metropolitan life. Even though his encounters in the city are often dangerous and filled with violence, his yearning for the metropolis separates him from the dogs.[72] Mamochka worries about his silent withdrawal from the pack and brings him "Puppy," a human baby who is alternately rival, playfellow, and brother to the young boy. That Mamochka is able to identify the boy's yearning for the company of other humans speaks to her remarkable empathy: like Dovey, Hornung is eager to show the mutual benefits of cross-species affection.[73] To Mamochka, if the child is a stray, he is to be taken in and fed, yet to most of the humans he encounters, Romochka is vermin, to be eradicated. There is a pathos to the way he is both drawn to and repelled by humans, given all that the novel tells us about their unreliable nature. This conflict plays out in the most horrific manner as the novel concludes, with Romochka signaling his choice to stay with Natalya and Dmitry by turning to the three puppies (Mamochka's final litter) he has cradled in his arms and fed from a milk bottle, and biting into their skulls. The work closes with the words "He had chosen to stay,"[74] suggesting that, for Romochka, conciliation between his two worlds is not possible. He must choose to be a human or a dog; he cannot be both.

Romochka ends the novel with a wild act, which contradictorily signals his commitment to the human, domestic world. However, the fact that he needs to make this choice at all suggests that he remains a stray. Creed explores the marginalization of the stray, calling for a "stray ethics": "A stray ethics focuses our attention on empathy for the marginalised and dispossessed. Compassion is crucial to relationships between human and nonhuman animals … A stray ethics asks us to put ourselves in the place of the other, while also recognising the importance of acting upon that experience through an ethics of care. The stray has the same rights as any other living creature to survive and flourish."[75] Both Dovey and Hornung depict the willed rupture of kinship ties and the loneliness and isolation that ensue when human beings deny or exploit interspecies connections. While Hornung's Romochka may ultimately reject the fluidity of boundaries between dogs and humans, he does so at enormous cost, and it is not clear that he will receive the support, care, and compassion that he needs within his newfound "family." His human status is steeped in animal death (a gesture loaded with macabre resonance) and, as such, it is a disturbing form of self-definition, grounded in notions of mastery and entitlement that are designed to discomfort the reader.

Freya Mathews articulates the ongoing ecological importance of embracing the nonhuman world. As she expresses it, "To engage with the unknowable subjectivities of animals, and to experience their response to us, is perhaps the principal bridge to communication with the unknowable subjectivity of the wider world."[76] Creed takes this idea a step further with her vision of "companion strays," recognizing a mutual dependence and balance that stem from shared custodianship of a (human) damaged world.[77] Through exposing the hoax of kinship, Dovey and Hornung challenge readers to reexamine their ties to animals—to hear their voices and to acknowledge their deep emotional capabilities. Animalography encourages a move toward such a system of ethics, telling the stories of those considered to be voiceless and insisting on creaturely empathy and inner lives, while also forcing us to confront our connectedness. We remain a long way from Herman's idea(l) of "co-authorship" or what Haraway terms an alertness to the "nonanthropomorphic competences of many kinds of animals,"[78] yet animalography plays an important role in reminding us to attend to the nonhuman other and to consider how we, and they, might communicate their lives in a more egalitarian fashion.

Dovey presents the voices of the dead to highlight the treacherous one-sidedness of human-animal relations, while Hornung posits an alternative system of care, emphasizing the tight loyalty of a canine pack in contrast to a failing human system. As Margaretta Jolly has commented, "Imagining the larger life story of the Chthulucene will condition our common survival in whatever form. A different way of understanding this is to ask what the dead are trying to tell us."[79] Both Dovey and Hornung show abuses and breakdowns of kinship, while their animalographies reveal the constancy of nonhumans in the face of human caprice. Dovey and Hornung reveal that there is much for us to learn from animal kinship networks, while they highlight the failures of established hierarchies. *Dog Boy* and *Only the Animals* both speak *for* animals, but they do so in a bid to encourage their readers to acknowledge them as kin, rather than eating them. They urge us to listen and to recognize the emotional sophistication of nonhumans. Furthermore, by encouraging us to understand ourselves as companion strays, both texts present alternative models for care and equality.

The University of Otago

Notes

1. Huff, "After Auto, after Bio," 279.
2. Bentham, *Introduction*, 311.
3. Creed, *Stray*, 100.
4. Huff and Haefner, "His Master's Voice," 153.
5. Haraway, *When Species Meet*, 97.
6. Mathews, "Living with Animals," 4.
7. For a discussion of the critic's sense of vulnerability when faced with the enigma of his cat, see Derrida, *The Animal*.
8. Huff, "After Auto, after Bio," 280.
9. Creed reminds us that animal refugees take flight from war, climate change, and other human-engineered phenomena, even though news reports are prone to concentrate on the plight of human casualties. See Creed, *Stray*, 71–75.
10. Hielscher, "A Joyous Deception."
11. Dovey, *Only the Animals*, 179.
12. Fudge, *Pets*, 10.
13. Herman, *Narratology beyond the Human*, 173.
14. Dovey, *Only the Animals*, 26, 20.
15. Dovey, *Only the Animals*, 204.
16. Dovey, *Only the Animals*, 206.
17. Dovey, *Only the Animals*, 206.

18. Disappointingly, Dovey's otherwise beautiful and moving recent study of J. M. Coetzee's work, *Ceridwen Dovey on J. M. Coetzee* (2018), does not engage with the novelist's representations of animals. Archer-Lean offers a detailed and engaging analysis of the connections Dovey weaves between her dolphin protagonist and Coetzee's academic character. As Archer-Lean asserts in "Revisiting the 'Problem,'" "Dovey's dolphin letter to Plath implies a push against the limit of realism as necessary to an empathetic animal representation." Archer-Lean continues to argue that Dovey is, through "textual self-referentiality and posthumous dialogue between constructed characters," affirming Elizabeth Costello's proposal that "we are all of one kind."
19. Dovey, *Only the Animals*, 207.
20. Bekoff, "Wild Justice," 75.
21. Bekoff, "Wild Justice," 75.
22. Bekoff, "Wild Justice," 73.
23. Herman, *Narratology beyond the Human*, 194. Herman follows Plumwood's assertion of anthropomorphism's utility at a time of environmental crisis. See Plumwood, "Journey."
24. Plumwood, "Journey," 33.
25. Huff, "After Auto, after Bio," 280.
26. Fudge, *Pets*, 5.
27. Fudge, *Pets*, 6.
28. See Plumwood, *Environmental Culture*, 45–61 and *Feminism.*
29. For a discussion of the significance of Franz Kafka's story to both Dovey and Coetzee, see Archer-Lean, "Revisiting the 'Problem.'"
30. Dovey, *Only the Animals*, 57.
31. Dovey, *Only the Animals*, 59.
32. Dovey, *Only the Animals*, 60.
33. Dovey, *Only the Animals*, 60.
34. Dovey, *Only the Animals*, 64.
35. Dovey, *Only the Animals*, 69–70.
36. Dovey, *Only the Animals*, 71.
37. Archer-Lean, "Revisiting the 'Problem.'"
38. Haraway, *When Species Meet*, 41.
39. Sutherland also stresses the close connections between humans and elephants in his biography of the Victorian show-elephant Jumbo. As he expresses it, "anthropomorphism [in relation to elephants] is not out of place. Pliny, the first historian to write at length about elephants, claimed that 'of all animals, the elephant is the closest to man in intelligence.' And closest to the man-child in its capacity to suffer. Many zoologists would agree." See Sutherland, *Jumbo*. Sutherland continues to draw attention to the elephant's capacity for grief, offering a graphic account of the sensory assault which the baby elephant that grew up to become Jumbo must have experienced through witnessing his mother's slow death. If this were not enough, parts of the mother elephant were then roasted and eaten, which Sutherland imagines in an appropriately grizzly manner.
40. Sutherland, *Jumbo*, 45.
41. Dovey, *Only the Animals*, 168.
42. Dovey, *Only the Animals*, 168–169.
43. Haraway, *When Species Meet*, 98. Dovey shows humans to be capricious and predatory, and what is all the more shocking is that the siege of Paris lasted

for only five months. As Baumgarthuber has noted in "Ill-Starred Lives," the city's wealthy managed to consume 220,000 sheep, 40,000 oxen, and 12,000 pigs before running out of animals that were bred to be food. The apparently staunch companion animals, dogs and cats, were eaten first, along with 70,000 horses, before the Parisian elite turned its attention to more exotic beasts, including the "pets of young Paris," Castor and Pollux, whose carcasses yielded 1500 kilograms of meat, which sold for sixty francs per kilogram.

44. Dovey, *Only the Animals*, 169–170.
45. Dovey, *Only the Animals*, 173.
46. Derrida, *The Animal*, 26.
47. Fudge, *Pets*, 18–19.
48. Fudge, *Pets*, 19.
49. Armstrong, *Sheep*, 50.
50. In his sensitive interpretation of the data generated by scientists to prove the sheep's powers of recognition, Armstrong notes, "Beyond the clinical language, we can clearly hear the voices of the sheep." See Armstrong, *Sheep*, 51.
51. Mathews, "Living with Animals," 9.
52. Mathews, "Living with Animals," 41.
53. Haraway, *When Species Meet*, 35, 36.
54. Hornung, *Dog Boy*, 11—although the narrator repeats this phrase throughout the novel whenever Romochka is faced with a choice.
55. Hornung, *Dog Boy*, 51.
56. Hornung, *Dog Boy*, 59.
57. Hornung, *Dog Boy*, 59.
58. Hornung, *Dog Boy*, 59.
59. Hornung, *Dog Boy*, 281.
60. Hornung, *Dog Boy*, 246.
61. Hornung, *Dog Boy*, 61.
62. Hornung, *Dog Boy*, 90.
63. Hornung, *Dog Boy*, 90.
64. Douglas, *Purity and Danger*, 44.
65. Hornung, *Dog Boy*, 81.
66. Hornung, *Dog Boy*, 131.
67. Rose, "How to Love."
68. Hornung, *Dog Boy*, 96.
69. Creed, *Stray*, 100–101.
70. Hornung, *Dog Boy*, 135.
71. Hornung, *Dog Boy*, 135.
72. There are other factors that prevent Romochka from completely assimilating with the pack, including the fact that his sense of smell is nowhere near as refined. While the dogs are able to track him, wherever in the city he may be, he is left helpless several times because he is unable to smell them or to find his way home.
73. Hornung depicts similar instances of interspecies empathy in her later novel, *The Last Garden*, most notably when Melba, a horse, grieves for her dead fowl, Lucie, and the human protagonist Benedict is "swamped" by both her misery and his own.
74. Hornung, *Dog Boy*, 81.
75. Creed, *Stray*, 100.

76. Mathews, "Living with Animals," 18.
77. Creed, *Stray*, 162.
78. Haraway, *When Species Meet*, 300.
79. Jolly, "Survival Writing," 481.

Acknowledgments

I thank Deirdre Coleman for introducing me to Dovey's writing, and Jessica White and Gillian Whitlock for suggesting Hornung as a companion author for this essay. Jess, in particular, made a number of extremely important suggestions which have, I hope, strengthened the piece, while Gillian made me aware of Archer-Lean's important article on Dovey, which appeared at the eleventh hour. I owe additional thanks to the anonymous reviewer, whose insightful suggestions were extremely helpful. My work on animals and emotions has been funded by the Australian Research Council's Centre of Excellence for the History of Emotions (project CE110001011), and I am deeply grateful to the Centre and its Chief Investigators for the support and opportunities I have received. I also thank my new colleagues at the University of Otago, where I completed most of the writing of this essay.

Disclosure Statement

No potential conflict of interest was reported by the author.

ORCID

Grace Moore http://orcid.org/0000-0001-5807-1475

Works Cited

Archer-Lean, Clare. "Revisiting the 'Problem' of Anthropomorphism through Ceridwen Dovey's *Only the Animals* (2014)." *Australian Literary Studies* 34, no. 1 (2019). https://www.australianliterarystudies.com.au/articles/revisiting-the-problem-of-anthropomorphism-through-ceridwen-doveys-only-the-animals-2014

Armstrong, Philip. *Sheep*. London: Reaktion, 2016.

Baumgarthuber, Christine. "Ill-Starred Lives." *The New Inquiry*, July 10, 2012. https://thenewinquiry.com/blog/ill-starred-lives/

Bekoff, Marc. "Wild Justice and Fair Play: Cooperation, Forgiveness, and Morality in Animals." In *The Animals Reader*, edited by Linda Kalof and Amy Fitzgerald, 72–90. Oxford: Berg, 2007.

Bentham, Jeremy. *An Introduction to the Principles and Morals of Legislation*. Oxford: Clarendon, 1823.

Creed, Barbara. *Stray: Human-Animal Ethics in the Anthropocene*. Sydney, NSW: Power, 2017.

Derrida, Jacques. *The Animal That Therefore I Am*. Translated and edited by David Wills. New York, NY: Fordham UP, 2008.

Douglas, Mary. *Purity and Danger: An Analysis of Concepts of Pollution and Taboo*. London: Routledge, 2005.

Dovey, Ceridwen. *Only the Animals*. London: Atlantic, 2014.

Fudge, Erica. *Pets (The Art of Living)*. London: Routledge, 2014.

Haraway, Donna J. *When Species Meet.* Minneapolis, MN: U of Minnesota P, 2008.

Herman, David. *Narratology beyond the Human: Story-Telling and Animal Life.* Oxford: Oxford UP, 2018.

Hielscher, Claire. "A Joyous Deception: Ceridwen Dovey's *Only the Animals.*" *Kill Your Darlings*, September 15, 2014. https://killyourdarlings.com.au/2014/09/a-joyous-deception-ceridwen-doveys-only-the-animals/

Hornung, Eva. *Dog Boy.* Melbourne, VIC: Text, 2009.

——. *The Last Garden.* Melbourne, VIC: Text, 2017.

Huff, Cynthia. "After Auto, after Bio: Posthumanism and Life Writing." *a/b: Auto/Biography Studies* 32, no. 2 (2017): 279–282.

Huff, Cynthia, and Joel Haefner. "His Master's Voice: Animalographies, Life Writing, and the Posthuman." *Biography* 35, no. 1 (2012): 153–169.

Jolly, Margaretta. "Survival Writing: Autobiography versus Primatology in the Conservation Diaries of Alison Jolly." *a/b: Auto/Biography Studies* 34, no. 3 (2019): 467–484.

Mathews, Freya. "Living with Animals." *Animal Issues* 1, no. 1 (1997): 4–20.

Plumwood, Val. *Environmental Culture: The Ecological Crisis of Reason.* London: Routledge, 2002.

——. *Feminism and the Mastery of Nature.* London: Routledge, 1993.

——. "Journey to the Heart of Stone." In *Creativity and Environment: New Environmentalist Criticism*, edited by Fiona Becket, 17–36. Leiden: Brill, 2007.

Rawson, Jane. "The Invisible Extinctions." *Meanjin* 77 no. 3 (2018): 54–62.

Rose, Deborah Bird. "How to Love a 'Pest.'" *Love at the Edge of Extinction*, February 21, 2014. https://webarchive.nla.gov.au/awa/20190725101447/http://pandora.nla.gov.au/pan/177305/20190725-1730/deborahbirdrose.com/2014/02/21/how-to-love-a-pest/index7067.html

Sutherland, John. *Jumbo: The Unauthorised Biography of a Victorian Sensation.* London: Anthem, 2014.

"Desperation for Life": Writing Death in the Anthropocene

By Jessica White and By Gillian Whitlock

A conservative analysis of extinctions of other-than-human species published in 2018 in the *Proceedings of the National Academy of Sciences of the United States of America* estimates that "as much as 50% of the number of animal individuals that once shared Earth with us are already gone, as are billions of populations."[1] As the extinction of populations is a precursor to species extinctions, "Earth's sixth mass extinction episode has proceeded further than most assume."[2] The authors estimate that "the window for effective action is very short, probably two or three decades at most" and that their research points "to ever more powerful assaults on biodiversity in the next two decades, painting a dismal picture of the future of life, including human life."[3]

Although these statistics center on wild animals, domestic animals are also suffering on massive scales.[4] Livestock (predominantly cattle and pigs) now constitute 60% of the world's mammals, while 36% are humans. The remaining 4% of mammals on this planet are wild animals.[5] Humans are breeding animals to kill them. Such onslaughts against other-than-human populations have not occurred since the asteroid impact at the end of the Cretaceous Period of the Mesozoic Era (about 65 million years ago), which wiped out 75% of all known species (including dinosaurs).[6] While the planet's five previous extinctions were caused by geological events, this time, as per the hallmark of the Anthropocene, humans are at the helm.

A life is given shape by its death and, as scholars whose trade is the writing of life, we cannot avoid the mounting deaths of other-than-humans that represent this era of "biological annihilation."[7] The three pieces in this forum contemplate ways in which literary criticism, autobiography, and informed imagination shape how we conceive of and examine the lives of other-than-humans such as cows, sheep, platypuses, and baiji, and engender consideration of and respect for their subjectivity and "desperation for life,"[8] to use Holloway's phrase from her essay in this section, against a background of death and extinction.

Jessica Holmes' "Writing the Cow: Poetry, Activism, and the Texts of Meat" is an innovative reading of poet Ariana Reines' collection *The Cow* that responds to the question, "What would it mean to 'write' the cow (not just to write *about* her)?".[9] Holmes draws on activist and scholar Carol J. Adams' *The Sexual Politics of Meat* to think through ways in which "*texts of meat* seek to erase, to abstract, and to decharacterize" animal subjects.[10] She proposes a "vegan poetics," which reacts to the backgrounding of animals in texts of meat. Rather than reducing animals to symbols, to something "sent to slaughter—translated, rewritten into a different kind of matter"[11]—a vegan poetics attempts to assert the cow's subjectivity and restore it to its living body.

Holmes meditates on the alignment between the silences of women and cows, and their transformation into objects. Barbara Holloway, in her essay "Sheep: Voice | Complicity | Precedent," ruminates on the correspondence between a lack of voice and agency between children and sheep. The cows in Reines' poetry become "the opposite of my body when they hang me up by my hind legs",[12] while the sheep in Holloway's essay are hung "upside down on the gambols from the rafter"[13] prior to being transformed from animal to "barely warm caryatid, with shortened, outstretched arms".[14] Both allusions signal a shift from subject to object, but where Reines uses poetry to challenge this process, Holloway draws on her autobiographical experiences.

When she was a child, Holloway's sheep-farming father brought home "poddy lambs" which had lost their mothers. The lambs formed "a mutually loving company with the children",[15] who bottle-fed them and lived with them as they grew. At the same time, each week a sheep was selected to be slaughtered, a process that Holloway was not permitted to watch, until she disobeyed orders. In an affective rendering of an event that Holloway "would do anything to unhear, to not know, for the rest of [her] life",[16] the animal is transformed, traumatically, from subject to object, from "the living sheep," an intra-active agent whose knitted wool kept Holloway warm, to "a cold body of meat".[17] This compels Holloway to ponder "the compromised relationship between human and flock animal".[18]

As the adult writing this essay, Holloway explores the contradictions of companionship and exploitation alongside E. O. Schlunke's *Stories of the Riverina*. His story "A Sheep's Eye View" allows a ewe to narrate her own life story, which, with its description of being "owned and handled by so many different men",[19] echoes the exploitation of female cattle in Reines' *The Cow*. "It is," Holloway writes, "a protest story; its metaplot condemns both cruelty and desire for profit beyond need",[20] recalling the role of the writer as an activist, using their craft as a call to action against killing.

Rick De Vos, in his essay "A Triumphal Entry, a Stifled Cry, a Hushed Retreat," deploys writing in a similar way, weaving perspectives of American specimen-collector Charles McCauley Hoy and his prey, two Australian platypuses and the (Chinese) baiji or dolphin, to prompt an awareness of the implications of the colonial fervor for hunting. Hoy, after three years in Australia from 1919–1921, during which time he hunted mammals to sell to the US National Museum of Natural History, noticed a decline in most native mammal species "due to a convergence of factors, including invasive species and land clearing".[21] These observations are eerie given that, a century later, research from fifty-nine scientists from across the globe indicates that the largest cause of the loss of wildlife is the destruction of natural habitats, predominantly for the creation of agriculture, and invasive species.[22]

The beauty of De Vos's piece is its positioning of viewpoints of the platypus and baiji, the journal entries of their hunter, and scholarly musing on the role of museums in creating what Australian scholars Penny Edmond and Hannah Stark describe as an "archive of bodies,"[23] or museums crammed with specimens collected from around the globe, particularly during the nineteenth century when novel (to Western eyes) flora and fauna from the colonies ignited a race for possession. The loss that accompanies the platypus's mate and the baiji's kin is contrasted with the US National Museum of Natural History's concerns that its collections would remain incomplete should these animals become extinct.

Each of these works draws on ecocritical scholar Kate Rigby's essay "Writing in the Anthropocene: Idle Chatter or Ecoprophetic Witness?" which dwells on the difficulty of ethical representation of the other-than-human in a world brimming with ecological disaster. Rigby, in turn, draws on Australian ecofeminist Val Plumwood's scholarship on the "logic of colonisation" or the notion that "instituting a rigid divide between subject and object, self and other, denies ethical considerability to anyone, or any thing, that is excluded from the company of the sovereign self."[24] The three writers in this forum destabilize this artificial boundary between object and subject by creating the voices of their subjects. In doing so, they mirror Rigby's observation that the plight of the other-than-human "obliges us to utterance, even though we know that our words, no matter how artfully wrought, are bound to be insufficient either to prevent or to bespeak the unprecedented horror of the ecocide of which we ourselves are both victims and perpetrators."[25]

Writing life and death means evoking absence and presence. Holmes refers to Carol Adams' concept of the "absent referent" in the context of industrial slaughter[26]—that is, the transformation of a presence into an absence through death and language; Holloway refers to the "spectral presence" of her father, so close to the sheep that they barely stirred "at his quiet walk as he stopped to help at difficult births"; while De Vos dwells on the poignancy of a death that is not just one death but a loss that ripples through an ecosystem. His baiji's words—"The world was different because of our presence and is different because of our absence"[27]—are a frightening reminder of the world that we are losing, both physically and figuratively. Unless we can stymie the loss, we may well become writers of death rather than of life.

The University of Queensland

Notes

1. Ceballos, Ehrlich, and Dirzo, "Biological Annihilation," E6095.
2. Ceballos, Ehrlich, and Dirzo, "Biological Annihilation," E6095.
3. Ceballos, Ehrlich, and Dirzo, "Biological Annihilation," E6095.
4. The distinction between "domestic" and "wild," or "native," in the Australian context is blurred. As Head and Muir note in "Nativeness," many writers posit 1788 (the date of British colonization) as a marker of ecological change, but Aboriginal people had been modifying their environment for thousands of years prior to this. For the purpose of this section of The Forum, "domestic" animals are defined, as per Hemsworth in "Farming of Animals," as those which have adapted to humans and the environment that humans have provided for the animals.
5. Bar-On, Phillips, and Milo, "Biomass Distribution," 6508.
6. Human activity did, however, contribute to the Quaternary Megafauna Extinction between approximately fifty thousand and three thousand years ago, which caused the loss of about half of the large mammal species. See Bar-On, Phillips, and Milo, "Biomass Distribution," 6508.
7. Ceballos, Ehrlich, and Dirzo, "Biological Annihilation," E6089.
8. Holloway, "Speaking As/For Sheep," XX.
9. Holmes, "Writing the Cow," 246.
10. Holmes, "Writing the Cow," 238.
11. Holmes, "Writing the Cow," 243.
12. Cited in Holmes, "Writing the Cow," 242.
13. Holloway, "Sheep," 254.
14. Holloway, "Sheep," 255.
15. Holloway, "Sheep," 254.
16. Holloway, "Sheep," 255.
17. Holloway, "Sheep," 255.
18. Holloway, "Sheep," 252.
19. Cited in Holloway, "Sheep," 257.
20. Holloway, "Sheep," 258.
21. De Vos, "A Triumphal Entry, a Stifled Cry, a Hushed Retreat," 265.
22. Grooten and Almond, *Living Planet Report 2018.*
23. Edmonds and Stark, "Friday Essay."

24. Rigby, “Writing in the Anthropocene,” 175.
25. Rigby, “Writing in the Anthropocene,” 175.
26. Holmes, "Writing the Cow," 239.
27. De Vos, "A Triumphal Entry, a Stifled Cry, a Hushed Retreat," 269.

Disclosure Statement

No potential conflict of interest was reported by the author(s).

ORCID

Jessica White http://orcid.org/0000-0002-9291-5712

Works Cited

Bar-On, Yinon M., Rob Phillips, and Ron Milo. “The Biomass Distribution on Earth.” *Proceedings of the National Academy of Sciences of the United States of America* 115, no. 25 (2018): 6506–6511.

Ceballos, Gerardo, Paul R. Ehrlich, and Rodolfo Dirzo. “Biological Annihilation via the Ongoing Sixth Mass Extinction Signaled by Vertebrate Population Losses and Declines.” *Proceedings of the National Academy of Sciences of the United States of America* 114, no. 30 (2017): E6089–E6096.

De Vos, Rick. "A Triumphal Entry, a Stifled Cry, a Hushed Retreat." *a/b: Auto/Biography Studies* 35, no. 1 (2020): 261–272.

Edmonds, Penny, and Hannah Stark. “Friday Essay: On the Trail of the London Thylacines.” *The Conversation*, April 5, 2018. https://theconversation.com/friday-essay-on-the-trail-of-the-london-thylacines-91473

Grooten, M., and R. E. A. Almond, eds. *Living Planet Report 2018: Aiming Higher*. Gland, Switzerland: World Wide Fund for Nature, 2018.

Head, Lesley, and Pat Muir. “Nativeness, Invasiveness, and Nation in Australian Plants.” *Geographical Review* 94, no. 2 (2004): 199–217.

Hemsworth, P. H. “Farming of Animals.” In *The Encyclopedia of Applied Animal Behavior and Welfare*, edited by Daniel Mills, Jeremy N. Marchant-Forde. Oxford: Centre for Agriculture and Bioscience International, 2010: 248–250.

Holloway, Barbara. “Sheep: Voice | Complicity | Precedent.” *A/b: Auto/Biography Studies* 35, no. 1 (2020): 251–260.

Holmes, Jessica. "Writing the Cow: Poetry, Activism, and the Texts of Meat." *a/b: Auto/Biography Studies* 35, no. 1 (2020): 237–250.

Rigby, Kate. “Writing in the Anthropocene: Idle Chatter or Ecoprophetic Witness?” *Australian Humanities Review*, no. 47 (2009): 173–187.

Writing the Cow: Poetry, Activism, and the Texts of Meat

By Jessica Holmes

> What are the words you do not have? What do you need to say? What are the tyrannies you swallow day by day and attempt to make your own, until you will sicken and die of them, still in silence.
>
> — Audre Lorde, *Sister Outsider*, 41

> The mouth's the haven for all an eye cannot disperse.
>
> — Ariana Reines, *The Cow*, 104

Meat-eating is the most prevalent way in which humans encounter animals. In 2018, over 33 million "cattle" were slaughtered for food in the US alone.[1] Despite the rise of vegetarianism and veganism, American meat consumption per capita is at an all-time high and continues to increase. The practice of consuming animal products, though habitual and frequently inadvertent in any "meaning" beyond satisfying personal appetite or taste, nevertheless constitutes participation in a social, cultural, and historical pattern of hierarchy and speciesism.

Activist-scholar Carol J. Adams introduced the notion of "texts of meat" in *The Sexual Politics of Meat*: "By speaking of the *texts of meat* we situate the production of meat's meaning within a political-cultural context. None of us chooses the meanings that constitute the texts of meat, we adhere to them. Because of the personal meaning meat has for those who consume it, we generally fail to see the social meanings that have actually predetermined the personal meaning. … [Meat] is seen as an item of food; its meaning recurs continuously at mealtimes, in advertisement, in conversations; and it is comprised of a system of relations having to do with food production, attitudes toward animals, and, by extension, acceptable violence toward them."[2] Adams' expanded notion of a "text" provides a useful framework by which to discuss the manifestation of such meanings in literature and, additionally, literature's potential to subvert those meanings. In order to begin to unpack the entangled relationship between humans and animals, and literature's role in it, this discussion piece will consider questions of anthropocentrism and representation in writing the lives of animals. More specifically, it will contemplate the writing of a cow's life. Through this contemplation, I introduce a *vegan poetics*, which I believe can help us to cultivate more ethically attuned reading and writing practices surrounding interspecies relations in the age of the Anthropocene. In her 2015 book *The Vegan Studies Project*, Laura Wright proposed "vegan studies" as a new area of scholarship. In my research, I have identified a need to define and demonstrate what such a framework might look like in the context of literary studies. Poetry in particular is a literary genre which has consistently lent itself to expressions of silenced or oppressed voices and bodies, in part due to its capacity to embody loss, fragmentation, and absence. Contemporary poems thus provide a useful foundation for rethinking narratives of anthropocentrism and revising discriminatory perceptions of oppressed or vulnerable bodies (both human and nonhuman). When engaged in such interventionary and reparative practices of thought, poetry constitutes a form of activism. Poems offer alternative methods of seeing or bearing witness to, remembering and assigning value to individual subjects.

Given life writing's connection to memory, experience, and personal testimony, *texts of meat* stand in direct opposition to life-writing genres, including poetry; through their normalized categorization of animal subjects, *texts of meat* seek to erase, to abstract, and to decharacterize. Ariana Reines' prize-winning poetry collection *The Cow* presents one of the starkest alternatives to a "text of meat," I argue, and exemplifies a productive (albeit unavoidably anthropocentric) intervention into narratives of violence dating back 2.5 million years.[3] Taking *The Cow* as my primary example, I ask readers of all types of "texts" what it is to represent an individual animal, whether an anthropocentric reproduction of a nonhuman animal can ever be accurate, and—if it cannot—how can translations of nonhuman animals in literature hope to advocate for the individuals they ultimately emblematize? What does it mean to "write" the life of a cow?

Adams' work emerges largely out of the "failure of feminists to recognize the gender issues embedded in the eating of animals."[4] Perhaps this failure is in part why Reines chooses the gendered figure of the (female) cow as her primary animal subject. The collection is as invested in the question of what it is to be female as it is in what it is to be cow. Reines' speaker(s)—fluctuating between distinctly human and vaguely animal—clearly identify a feeling of kindredness between woman and cow that emerges predominantly out of shared experiences of suffering. Within industrial food systems, most male calves are killed shortly after birth (though some are raised for leather or to supply semen), whereas dairy cows are kept alive for their reproductive capabilities (to supply milk). Both the cows and women in Reines' poems are extremely sexualized and objectified, and frequently constitute victims of violence. They are also defined repeatedly by a perceived bodily absence; their bodies serve as holes to be penetrated or vessels to be emptied, filled, and emptied again. The collection includes many graphic scenes of anal penetration, intercourse, and oral sex, but the vision of the female body as vessel or absence is not purely sexual. In "In Which She Pays for Her Tardiness," Reines writes:

> There was a kink in the thing that was a girl
>
> ...
>
> I was a sock filled with rubble
>
> ...
>
> I was the shaft some light filled
>
> I was a skin
>
> They filled me with something
>
> I was a LUNG.[5]

To be a "girl" is to be a "thing," a container to be occupied. The container is sometimes occupied by blood, semen, antibiotics, rubble, light, flesh, or air.

Throughout *The Cow*, Reines lends a presence to this amorphous absence. One reason I deem *The Sexual Politics of Meat* such a generative framework through which to examine *The Cow* is because of Reines' masterful capacity to navigate the oscillation between absence and presence in the bodies of her poetic subjects. In *The Sexual Politics of Meat*, Adams applies the literary term "absent referent" to animal subjects in the food industries.[6] Through slaughter, animals become absent referents, transformed from live individuals into "food." According to Adams, they are made absent literally (through death), definitionally (through terminology), and metaphorically (through imaginative repurposing).[7] Poets without a doubt participate in the second and third modes through their deployment of animal subjects; Reines is no exception. She problematizes the very act of representation in "Lodge": "There is a stupidity in the conflation I am in the act of, cow with cattle car and mother with me, cunt and carcass and book and stomach. But this stupidity, if it belongs to me, is also exterior to me. Humans got brutalized by being packed into cattle cars and dying in them or by them which in turn humanizes, necessarily, the suffering of the beasts for which cattle cars were made. Then what. Signification is incestuous, iterative, autofellating."[8] To refer to the animal "holocaust" or the "rape" of cows is contentious in part because, as Adams puts it, "some terms are so powerfully specific to one group's oppression that their appropriation"—despite very literal parallels such as the slaughter of pigs in gas chambers or the forcible impregnation of dairy cows—"is potentially exploitative."[9] Poetry's reliance on signification and metaphor is potentially problematic in this way. In her article "Writing in the Anthropocene," Kate Rigby makes a case for the value of "writing into the Anthropocene in the mode of prophetic witness," while also acknowledging that in a context of modern ecocide, language runs the risk of amounting to "little more than 'idle chatter.'"[10] *The Cow* focuses more on presence than prophecy, but, nevertheless, by participating in a mode of life writing that bears witness to violences of such intensity and scale, Reines tests language's capacity to walk a difficult line–between the "idle chatter" Rigby describes and the kind of potentially exploitative terminology to which Adams refers. Rigby situates her argument for ecological ethical consideration within the gap between "sovereign self" and other: "the relative unconcern—the refusal to witness—with which the fate of nonhuman others is generally regarded, notwithstanding the efforts of individuals and minority groups, whether that fate is traced in the extinction of whole species or in the day-to-day suffering and death of animals in the meat trade, in laboratories, in recreational hunting, or simply on our blood-spattered roads, holds a clue as to the structural connection—but not, I repeat, equivalence—between genocide and ecocide."[11]

In an animal-rights context, even if we refrain from using terms such as "holocaust" and "rape" that risk rendering different violences as equivalences, to conjure the voice or the perspective of a cow in a poetic context is to anthropomorphize, to participate in the imaginative rewriting of a nonhuman individual's ontological experience. Even as Reines acknowledges a "stupidity" in the signification in which she engages, such implementation of exploitative, "incestuous" signification through poetry is "also exterior" because such practices manifest themselves in our institutions, our cultural traditions, our language, and our history at large. One does not have to be a poet to engage in and reproduce "life writing." But by linking the multitudinous and overlapping erasures of women and cows in her poems, Reines not only exposes what Adams refers to as the institutionalized, patriarchal—and I add *speciesist*—"structure of the absent referent,"[12] but ultimately ignites a long and complicated process of restoring that referent by lending a voice to that which is silent, a face to she who is anonymous, and presence to all that is absent. "There are vessels that have had their innards emptied out," Reines writes definitively in "Dear Marguerite."[13] In "Leftovers," however, her speaker searches for "the residue of what a body is doing but somehow in the iterative efflux of sound in which meaning resonates containing also the residue of others who are dead, so that the substance of signification is also ghosts."[14] *The Cow*'s mode of prophetic witness looks to the meaning resonating from disappeared and othered beings. Through representation, the absent referent(s) are there and not there, ghosts the poet refuses to let disappear completely but inevitably fails to bring back to life.

When animal-rights activists participate in the practice of "bearing witness" to animal suffering and slaughter, a commonly stated objective is to see animals for what—or rather *who*—they are, as opposed to commodities, property or future food products, and usually with the hope of experiencing some kind of empathy or connection. Just as Reines' representation of "ghosts" walks the line between absence and presence, the practice of bearing witness likewise fails to bring animals back to life (or to prevent their deaths in the first place). It is a symbolic act in many (though not all) ways, but one in which participants actively rewrite dominant portrayals of animal bodies through their own embodied experience (that of witness). Accordingly, a vegan poetics does not pretend to do away with all metaphor—to do so would be impossible, and some metaphors are indeed quite productive (more on this later)—but a vegan poetics does sustain an allegiance to literal animal and human presence, to the singular beings and bodies upon which the human gaze falls.

How is one to read "the cow" in the context of a poem? Adams states that a major challenge of vegetarian (and, I argue, even more so vegan)[15] activism is "being heard about literal matters in a society that favors symbolic thinking."[16] Poetry likewise has a tendency to favor symbolic thinking. One could feasibly choose to read every mention of "cow" in Reines' collection as a metaphorical stand-in for "woman." I argue, however, that the book resists this kind of purely allegorical reading or anthropomorphic allegiance. Reines routinely juxtaposes the lyrical with the literal, particularly during the nineteen-page series entitled "Rendered" that occurs late in the collection—a series in which she details various processes of and materials involved in cattle slaughter in exceedingly accurate detail.[17] Alongside an undeniable lyrical dynamism, Reines describes various encounters of violence (toward both cows and women) with graphic precision, maintaining an allegiance to the literal that works to thwart the erasure of violence and oppression—including erasures carried out through metaphor. The literal (no-longer-absent) referent takes up space in the poems. Or, as Adams puts it, "[t]o make the absent referent present—that is, describing exactly how an animal dies, kicking, screaming, and is fragmented—disables the power of metaphor."[18] A vegan poetics calls on the reader to contemplate the cow herself, not her absence.

For all the compelling factual details Reines provides, when it comes to the cow's perspective, the reader might ask: How literal can a poem be? And why study poetry as opposed to a more traditionally literal genre such as nonfiction, documentary, or journalism? The first direct reference to the body of the cow occurs in "Blowhole" early in the collection: "Moistest mouth is cow's mouth sorrow face normal. Hung up hind legs reversed negated shit brains rear world."[19] In literal terms, this is a description of a cow about to be or having just been slaughtered—"[h]ung up" by its "hind legs." (In a standard slaughterhouse practice, cows are hung upside down so as to have their throats slit and then to bleed out.) A few pages later, in "Knocker," we get the same scene, this time from the supposed cow's perspective: "My body is the opposite of my body when they hang me up by my hind legs."[20] Thus, even as Reines moves into the territory of anthropomorphosis or projection of human voice onto the nonhuman body, she builds on the literal. And as "they" hang her up, a rewriting of subjecthood occurs; reversal and negation serve as forms of erasure, forcibly stamping out the animal's essence to reveal "shit brains rear world." Literally (because she is upside down), the animal's bottom or "rear" has replaced her face, the primary outward expression of identity and suffering (that suffering already normalized in the eyes of the beholder—"sorrow face normal"). The language also echoes *rear* as in "to raise" or "to bring up," as a mother rears her child, and, in this way, Reines gestures further toward the notion of an upside-down or despicable world brought forth not by any motherly body but by "shit brains."

Adams condemns our society's "failure to acknowledge the importance of body-mediated knowledge."[21] Meat serves as a kind of antithesis to the living body; as one awakens to the "revelation of the nothingness of meat," "one sees that it came from something, or rather someone, and it has been made into a no-thing, no-body."[22] *Texts of meat* perform the opposite of life writing. Presence is translated—violently—into absence, "she" into "it," subject into object. The body of the mother specifically endures this translation repeatedly in *The Cow*. In "In the Most Holy Place Shalt Thou Eat It," Reines writes, "To be a blank upon which the hells project their sorrow and to forgive them, that means to be a mother."[23] In the case of the mother, the language reflects in several instances a kind of unsung nobility in such emptiness, a nobility being desecrated with each transliteration of the body—through oppressive forces such as violence, enslavement, sickness, time, and (eventually) death. In "Dear Marguerite," Reines writes, "Time's up for the mommas time's up for the mommas they're all gonna get too old for longer and longer and die slower and slower and you're not gonna care, are you, gentle reader."[24]

A vegan poetics asks us to care, pushing back against any purported tension between feeling and aesthetics. The mother serves as the ultimate vessel for absence|presence, as her body literally carries her child, is emptied of that child, fills up with milk, then is emptied of that milk. Or, in the case of industrial dairy cows, the body of the mother is drained dry not by her child but by the milking machine, re-impregnated, drained dry, re-impregnated, drained dry—"Delve into the marzipan of her rutted teats. | You have got to squeeze her squeeze her squeeze her," reads "Valve"[25]—and so on, until her body is spent and thus economically devoid of value. And so she is sent to slaughter—translated, rewritten into a different kind of matter.

As it continuously undergoes rewriting, the body quickly becomes detached from initial notions of vitality, mobility, and freedom. In "Blood Libel," Reines starkly narrates, "Then you find out that the gift of a body alone doesn't guarantee any kind of life."[26] For the cows, for the women, and for the speaker herself, the body seems to be most closely associated with pain—such as in "Dear Marguerite: "Body is whole, body hurts. Body is whole, body hurts"[27]—and loss or absence of selfhood—such as in "Rendered":

> Where am I not in my body, where am I least in it. What could be excised from me most easily.
>
> Where is a living thing not itself. Is her shit any less her.
>
> ...
>
> Where does life exist.[28]

This series of questions suggests a grasping for the kind of body-mediated knowledge that Adams believes has long been ignored. In a world where dominant practices of seeing and speaking repeatedly advance the erasure of suffering, the speaker in "Advertisement" wonders, "Why cannot a body itself be testimony."[29] In the animal-activist examples of vigils, acts of bearing witness, and undercover footage, the animals themselves become just that—testimony or proof of their lived experiences and the conditions of their slaughter. However, the controversial nature of such practices and, in many cases, the legal regulations they brush up against contend that, in the eyes of most humans, a body does not constitute testimony. *The Cow* importantly aligns itself with activist practices by setting its sights on the restoration of the referent by means of the body—that is, testimony of "life" lived, "a living thing." Of course, just as not all bodies are deemed equally worthy of life, not all language is deemed equally valuable. Poetry, like legally sanctioned speech or "testimony," has undergone methods of framing and forming that separate it from everyday language. Part of the difficulty of restoring the absent referent through poetry lies in the limitations of language, as well as the danger in metaphor.

A vegan poetics rethinks the status of metaphor within poetic utterance, seeking to link ethical commitments more directly to form. At the back of *The Cow*, Reines includes an image of two naked women with oversized breasts, heaving pregnant bellies, udders, and horns. The image visually illustrates the linked experiences of women and cows, but also draws attention to the ridiculousness of such a metaphor. The image, unaccompanied by any text or specified origin, exudes a certain comical perversion and reminds us that all representation carried out by humans is anthropocentric at heart. In "Dear Marguerite," Reines writes matter-of-factly, "DEAR WOMEN | I know that you are not cattle."[30] For all the conflation between women and cows on which the book hinges, it asks—of the reader and perhaps of the writer as well—that this conflation be a restorative measure, a means to expose oppression and suffering, not to hide or advance them. According to Adams, feminism questions the structure of stories, locating feminist meanings in the "gaps and silences of traditional narrative."[31] A vegan poetics aligns itself with feminism in this way. The literal silence of the cows—both in the sense that they cannot speak words and in the silence of death—occupies space in many of Reines' poems, haunting the abundant white space, even as it gives way to anthropomorphic revisions: "Eat me. | Eat me. | Silence of eternity," ends the poem "Decorticated."[32] No cow wants, let alone begs, to be eaten. The human voice attributed, however facetiously, to her reminds us that a fundamental distinction between the woman and the cow is that the woman (speaker|writer-figure included) possesses the tools of words while, from the perspective of the human, the cow has none and thus in one respect remains voiceless, silent. Farmed animals are, in fact, far from voiceless or silent, communicating through bodily utterances such as lows and cries, but such "language" is rarely listened to or valued in the setting of an industrial kill facility. Though the speaker repeatedly expresses a feeling of futility in language, when reading|writing the life of the cow, we must consider how far from the wordless animal it is appropriate to place the figure of the woman, given her possible empowerment through voice. With each poem, the speaker renews her power of lyrical expression; this expression contrasts with the haunting permanence of the eternal silence associated with the "decorticated" or skinned animal body, which will be transported, processed, and packaged into someone somewhere's meal: "Eat me. | Eat me." The quiet that follows seems not just to probe the gaps and silences of the traditional narrative of *texts of meat*, but attempts to rewrite that narrative altogether.

To write about cows inevitably involves representation; *texts of meat* argue that such beings are not of our species and are thus different from us, disconnected, "other." In "The Seed Is Rotten Under Their Clods," the speaker claims, "We don't care what the fucker feels. Sensation that is something wholly other. Long night of guts."[33] While *The Cow* does not go so far as to argue that the woman's perspective is the cow's or that the cow is entirely representative of the woman, the repeated conflation of bodies throughout Reines' poems resists this notion of complete otherness. What would a sensation that is "wholly other" look or feel like, after all? How would one write about it? While they are alive, the bodies of the cows have faces, exhibit emotions, have relationships, are violated, feel suffering, cry out—"Acres of wishes inside of her," reads "Knocker."[34] Is it not more likely that in order to endure the "Long night of guts," to murder and decorticate living beings, or even simply to witness such an act, one must *make* their bodies other—as the slave master does to his slave, as the rapist to his victim, as the man does who, in "In Which She Pays for Her Tardiness," hurls a tungsten lamp at the woman he claims to love? The speaker of "Se trouver dans un trou, au fond d'un trou" professes,

> When the difference between seeing and others, and all else, becomes clear as a veil.
>
> I start to know something that is a membrane somebody licked onto my two eyes.
>
> …
>
> Is this shithole my membrane, do I have to feel the ends of what I feel by virtue of what it is.[35]

This notion of a "veil" or "membrane" questions the very nature of seeing, of the individual human's capacity to bear witness. It impedes visibility and obscures its true subject, enabling the viewer to *make other*. The speaker questions the ends or limits of individual feeling and the compulsion behind those limits; perhaps "[w]e don't care what the fucker feels" "by virtue of what [not *who*] it is"—a cow. What if sensation only looks *otherly* on account of some "shithole" membrane or veil, not even one's own?

In a later poem, "Trope," we encounter yet another representation of obscured reality:

> Now that my face is become a hood
>
> Over the pulp of bad stuff that daily harms its outlook. Now that I know
>
> That it is not for them I cry but for me. For me.[36]

This admission of human selfishness pervades the collection, bringing into question the (un)ethical impulses often caught up in acts of seeing and writing other beings. What would it mean to "write" the cow (not just to write *about* her)? Reines clearly suggests that a translation or altering occurs through language (words are a human construct, after all), but wants to provide an alternative "writing" of these living bodies to the ones performed by *texts of meat*, in books, and in slaughterhouses and restaurants alike. In "Billet," she writes, "I cannot count the altering that happens in the very large rooms that are the guts of her. | Words translated."[37] What kinds of translations and representations of a body are possible? And might we seek out a less exploitative (re)writing, one that is truer to the cows themselves, despite the impossibility of really knowing them? In *The Sexual Politics of Meat*, Adams recalls eating a hamburger as a child while mourning the death of her pony: "I could summon no ethical defense for a favoritism that would exclude the cow from my concern because I had not known her. I now saw meat differently."[38] In this moment, a politics of grief is suddenly unconcealed from Adams' view—one in which degrees of mourning directly correlate to degrees of personal loss, and the disparity between the cow's and the pony's personal value exists only on account of a deeply unethical "favoritism" (in this case, what animal-rights advocates and scholars call "speciesism"). A vegan poetics attempts to lesson this disparity, shrinking the distance between an abstract or "other" being and one which is valued because known. Reines' poems attempt to "see differently" by challenging us to consider what it means to "know" an animal—to read her, write her, rewrite her, represent her, anthropomorphize her, use her, consume her.

What is the relationship between poetry and activism? At times, Reines' poet-speaker sounds empowered in her defiance of traditional narratives of oppression: "I will not train myself to love this shit," declares the speaker of "Dear Marguerite."[39] However, she also acknowledges the potential uselessness of writing as a form of activism or change: "What does a country need a poet for. To put up bunting on the dead shanks of dreams," asks "Valve."[40] And, in "Blood Libel," the speaker notes how "Everything can be put to use except the low. The utterance. | The language of meat doesn't change much."[41] In this way, Reines aligns the cries of the cows ("lowing") and poetry in their alleged uselessness. Adams would surely agree that "[t]he language of meat doesn't change much" as we witness its meanings getting reproduced over and over again in our lives. If we are to stop replicating cruel, inhumane practices toward humans and nonhumans alike, we must stop reproducing their meanings. And in order to do so, we must attempt to remove the "membranes" from our eyes, the "veils" and "hoods" from our faces. We must not "*dis*member the meaning of meat from the animals' lives" but must instead *re*member the live body, the referent, the individual.[42] A vegan poetics marries poetry to life, memory, and experience, however fleeting. Reines seems to find value in the rewriting of *texts of meat*, even in moments of futility. In "Blood Libel," she writes,

> The language of meat doesn't change much. But still if there is something I can do to shit reality by writing I have to.
>
> ...
>
> I have to find out how writing has to do with what happens.[43]

Adams offers one clue as to writing's relationship to activism, contending that "[v]egetarianism is an act of the imagination. It reflects an ability to imagine alternatives to the texts of meat."[44] Poetry, I argue, is aligned with vegetarianism and veganism in this way, in its imaginative capacity to transform a body. And Reines demonstrates through her collection how a writer might make the most of such an alignment by making remembrance (*not* dismemberment) a literary aspiration and priority.

The Cow seems to ask: What can a book do? What agency might be awakened by foregrounding the remembrance of animal bodies? In "A Cleaner, Safer World," Reines writes, "Can a book carry you into the world you have to pretend doesn't exist most of the time, can a book carry you back out into what first made you alive. Muscles in their marbling enfolding the sickest livers and the saddest eyes, and tongues that flop out pinking as they dessicate."[45] Furthermore, I ask, can a book (such as this one) lift a veil or membrane to reveal a living, suffering body? Can it connect a reader to "what first made you alive"—matter, body, mother? Can a book be a vehicle that "carries you" toward a different, more compassionate kind of seeing? Adams describes *texts of meat* as "structures that transform flesh ... into meat ... [and] signal the degree of distancing that our culture has determined is necessary for consumption of animals to proceed."[46] She stresses the need for alternative cultural interventions and interruptions from all directions, including literature. If texts can transform flesh in this way, I wonder, can texts also transform it back, reverse it, unerase individual subjecthoods? Can Reines, for example, make us see the cow instead of the meat?

An undercurrent of numb guilt pervades much of the collection—"I am too sorry to know how I feel," admits the speaker in "Please Be the Beef, Please Beef, Pleasure Is Not Wailing. Please Beef, Please Be Carved Clear, Please Be a Case of Consideration."[47] The speaker's desire to help both the women and the cows is dependent on recovering sensation. In "Rendered," Reines calls for a sweeping reestablishment of feeling: "I want to found a country where everybody feels. | Universes shooting out of matter so tiny you can only feel it."[48] Where seeing is flawed or thwarted by forces of erasure, feeling holds the capacity for a different kind of "knowing." However futile language may prove for Reines and|or her speaker(s), *The Cow* certainly rouses such feelings in its reader—if not emotional sensations such as sadness or sympathy, then visceral ones at the very least. In its graphic descriptions of blood, mucus, and feces, for example, it succeeds in spoiling one's appetite. In this respect, while not all poetry can (or aims to) argue on behalf of animals or oppressed bodies at large, it can help us relearn how to feel—sorrow, disgust, anger, connection, empathy.

"Item" opens with a reference to Hindu impressions of the cow: "It is not that you have to worship her. You just have to not kill her."[49] While the vegan reverberations of *The Cow* may seem radical to the everyday reader (and the everyday meat-eater), Reines' effort to displace and disarm *texts of meat* through poetry instigates the most basic of reparations. Instead of a *text of meat*, she offers up a *text of cow*. The book is, after all, entitled simply *The Cow*. By choosing to title the collection thus, she reaffirms the animal itself, the living body, not its translation. According to Adams, being "in touch" with the body "restores the absent referent and body-mediated knowledge."[50] Reines certainly proves to be "in touch," but through *The Cow* she also makes clear how complex, incomplete, and precarious this process of restoration remains. There is more work to be done. The literary community must join the likes of Adams and Reines in more directly pursuing a vegan poetics. Poetry constitutes a form of activism, by which I mean it *actively* strives against the ongoing erasure of oppressed bodies. When we "make other"—through the use of language and|or the consumption of that language—we act as poor readers and poor writers. And our humanism is a poor excuse for our inhumanity. When we strive to rewrite *texts of meat* and to engage in more compassionate, truthful practices of "knowing," we also learn to be better readers and writers, better literary critics, and better critics of the world at large.

University of Washington

Notes

1. United States Department of Agriculture, "Livestock."
2. Adams, *Sexual Politics of Meat*, 24.
3. It is estimated that human ancestors began regularly consuming meat about 2.5 million years ago. Mayell, "Evolving to Eat Mush."
4. Adams, *Sexual Politics of Meat*, 26.
5. Reines, *The Cow*, 26.
6. For a lengthier explanation of the "absent referent," as well as additional examples, see Adams, *Sexual Politics of Meat*, 51–56. Adams refers to Margaret Homans' discussion of the absent referent in literature in *Bearing the Word*, 4.
7. Adams, *Sexual Politics of Meat*, 52–53.

8. Reines, *The Cow*, 100.
9. Adams, *Sexual Politics of Meat*, 54.
10. Rigby, "Writing in the Anthropocene," 173, 174.
11. Rigby, "Writing in the Anthropocene," 175.
12. Adams, *Sexual Politics of Meat*, 53.
13. Reines, *The Cow*, 42.
14. Reines, *The Cow*, 89.
15. While Adams refers primarily to vegetarianism, I believe her vegetarian-feminist critical theory can and should logically be extended to include and argue in favor of veganism.
16. Adams, *Sexual Politics of Meat*, 85.
17. Reines, *The Cow*, 65–83. Examples include explanations of the "Z value" (the temperature increase required to reduce thermal death), "downer cows" (mature dairy cows that remain recumbent hours after calving), vaginal and anal explorations of cows, and various specific steps in standard meat processing.
18. Adams, *Sexual Politics of Meat*, 63.
19. Reines, *The Cow*, 4.
20. Reines, *The Cow*, 11.
21. Adams, *Sexual Politics of Meat*, 167.
22. Adams, *Sexual Politics of Meat*, 187.
23. Reines, *The Cow*, 38.
24. Reines, *The Cow*, 42.
25. Reines, *The Cow*, 54.
26. Reines, *The Cow*, 103.
27. Reines, *The Cow*, 42.
28. Reines, *The Cow*, 83.
29. Reines, *The Cow*, 52.
30. Reines, *The Cow*, 41.
31. Adams, *Sexual Politics of Meat*, 105.
32. Reines, *The Cow*, 6.
33. Reines, *The Cow*, 8.
34. Reines, *The Cow*, 11.
35. Reines, *The Cow*, 10.
36. Reines, *The Cow*, 53.
37. Reines, *The Cow*, 16.
38. Adams, *Sexual Politics of Meat*, 12.
39. Reines, *The Cow*, 43.
40. Reines, *The Cow*, 54.
41. Reines, *The Cow*, 103.
42. Adams, *Sexual Politics of Meat*, 165; emphasis added.
43. Reines, *The Cow*, 103.
44. Adams, *Sexual Politics of Meat*, 191.
45. Reines, *The Cow*, 56.
46. Adams, *Sexual Politics of Meat*, 163.
47. Reines, *The Cow*, 51.
48. Reines, *The Cow*, 83.
49. Reines, *The Cow*, 31.
50. Adams, *Sexual Politics of Meat*, 177.

Disclosure Statement

No potential conflict of interest was reported by the author.

Works Cited

Adams, Carol J. *The Sexual Politics of Meat: A Feminist-Vegetarian Critical Theory*. New York, NY: Continuum, 1990.

Homans, Margaret. *Bearing the World: Language and Female Experience in Nineteenth-Century Women's Writing*. Chicago: University of Chicago Press, 1986.

Lorde, Audre. *Sister Outsider: Essays and Speeches*. Trumansburg, NY: Crossing, 1984.

Mayell, Hillary. "'Evolving to Eat Mush': How Meat Changed Our Bodies." *National Geographic*, February 18, 2005. https://www.nationalgeographic.com/news/2005/2/evolving-to-eat-mush-how-meat-changed-our-bodies/

Reines, Ariana. *The Cow*. New York, NY: Fence, 2006.

Rigby, Kate. "Writing in the Anthropocene: Idle Chatter or Ecoprophetic Witness?" *Australian Humanities Review* 47 (2009): 173–187. http://australian-humanitiesreview.org/2009/11/01/writing-in-the-anthropocene-idle-chatter-or-ecoprophetic-witness/

United States Department of Agriculture. "Livestock and Meat Domestic Data." Accessed November 9, 2019. https://www.ers.usda.gov/data-products/livestock-meat-domestic-data/

Wright, Laura. *The Vegan Studies Project: Food, Animals, and Gender in the Age of Terror*. Athens, GA: U of Georgia P, 2015.

Sheep: Voice | Complicity | Precedent

By Barbara Holloway

> Is it not first through the voice that one becomes animal?
>
> — Gilles Deleuze and Felix Guattari, *A Thousand Plateaus: Capitalism and Schizophrenia*

It is strange, but I do not remember an individual sheep's name. I remember Humphrey, Sluggo, Ike, Peggy, Jinna—horses and cows, dogs and birds—but not one sheep. I should remember: from our first moments, for my family, people lives and sheep lives were daily intertwined. Poddy lambs shared the kitchen. The grown sheep called out to each other as they grazed past the house. We had the feel of their coats on our fingers—short and curly lambs; smooth, grey outer tips on grown sheep with the soft, cream crimp wool below. We breathed the warm, mellow presence of sheep, their sheepy grass-fresh droppings. The human talk and movement were made up of the sheep, their needs, and the associated rituals. My mother knitted woolen singlets and jumpers. We wore wool hats and coats. Their dried manure grew our food. And their flesh became our flesh.

It is so long ago that not a single sheep would remember me, but even now, walking past sheep in a paddock anywhere, I cannot help calling out to them. Some of them call across the paddock. I answer. Some look up then. They stop grazing, they come over or call to me, and we look at each other. … I know different pitches have different significance, indicating warning, well-being, or friendliness, but still it is not language, theirs or mine.

Language is a different matter to voice. "[A]lways [the animal's] lack of common language, its silence, guarantees its distance, its distinctness, its exclusion from and of man [*sic*]," wrote John Berger in "Why Look at Animals?"[1] This essay's key themes are animals, language, and relationship: "With their parallel lives, animals offer man [*sic*] a companionship which is different from any offered by human exchange."[2] Berger's many examples of the intertwining of human and animal lives in human cultures, even in the fundamentals of human language, show how complex and multifaceted the connection is. He does not discuss these in the context of the exploitation of domesticated animals. The discussion of relations between human and other-than-human in writing has developed radically since his essay in the context of the dire predicament we know as the Anthropocene, refiguring the status of animals, their treatment by humans, and representation of their lives in literature. Nonetheless, it provides one starting point for exploring relations between sheep and humans read against new materialism's different perspectives on being, its renewed understandings of self and subjectivity, of human, and "the myriad of other beings that are our kin."[3]

Of that myriad, my essay is an investigation of literary forms of voice in the compromised relationship between human and flock animal. It investigates the possibility of other-than-human agency in writing. I first revisit and attempt to unpack formative relationships I had with sheep in order to address issues embedded in interdependency and interactivity between human and pastoral animal. Such relationships are hard to articulate, so innate do they seem in my being—intuitive, complex, emotional knowledge beyond thought. Drawing on the lens of critical posthumanism and alert to the (post)-colonial context, I try to engage with inherent and simultaneous companionship and exploitation, ministry, empathy, and ruthlessness, exemplified by both innocence and complicity, as I recall induction into that companionship. I go on to consider a portrayal in earlier literature of such ambiguous relations, revaluing it in the context of separation from animals in contemporary urban society.

***

Cynthia Huff refines the materialist argument by envisaging a "critical posthumanism" and a "posthumanist imaginary" that can decenter the human in writing. That imaginary would "deprivilege sight—the sense with which humans most identify and the one they most use—and qualitative cognition by equally favoring touch, taste, smell, and hearing as well as quantitative thinking, all ways of knowing the world that are just as or more important for non-human animal and mechanical reality."[4] But importantly, a posthumanist imaginary would also be alert to the context of the environment. It would "consider the surround in which beings interact."[5] That "surround" makes it easier to decenter the human by bringing the other-than-human to equal status in auto|biography. It is important in a post-colonial environment in which introduced animals — especially the livestock of pastoralism — play a major part in the "surround", by environmental damage, endangering native species and undoing the ecological balances achieved by Aboriginal culture and practice.

Although premodernist precedents lacked the values and concerns of the present, they have the advantage of creation from familiarity with sheep and other beings. By contrast, Philip Armstrong, author of the monograph *Sheep*, reports that he met a living sheep only by going to a pet farm as part of his research.[6] The realities of modernism, urbanization, and industrialization have divorced humans from other lives, ending cultural and physical familiarity with animals.[7] We cannot know how much writing in the modern period was rejected for publication because it presented a multitude of beings, creating their environment instead of tracing the life of the "proper autobiographical subject, a representative man," in which other beings are present only to trace out his narrative.[8]

A spare Australian literature exists with an awareness of such "intra-acting"—Huff's term. In 1931 Frank Dalby Davison published *Man-shy*, the life of a heifer written from her point of view, and in 1940, *Dusty*, the life of a kelpie-dingo cross.[9,10] More recently, some of Gillian Meares' fiction could be described as intra-active, as in "The Whippet and the Willow," which begins, "Emily Egan lies underneath the rug made out of six hundred and thirty-six hexagons she crocheted when the whippet was no more than a puppy. He is an old dog now and the rug smells of his skin and her skin, or of an intermingling of the two, one indistinguishable from the other, but not at all unpleasant."[11]

A suite of stories by E. O. Schlunke set in the New South Wales Riverina region in the mid-twentieth century is a more comprehensive though earlier exception. As a regional writer, Schlunke created a literature in which the human is only one participant among innumerable interdependent agents, including sheep. Those agents also contribute to the crafting of his texts. I will introduce some details of his work later, but first write an aspect of my own relations within a sheep culture.

On the Edge of the Central Tablelands, Lambing Time, August-September

The shepherd, my father, gets up before light every morning, whatever the weather this time of year. No dogs, no children nor his wife stir out of their sleep. He puts on an aging army greatcoat, scarf, hat, and gloves, and walks out in the frost and fog. His feet know the track until the first gray light spreads across the grass when he reaches the ewes and lambs. It needs to be that hour, the time of birth and death, the time when foxes creep into the flock to kill and eat newborn lambs.

These were discouraged by his spectral presence, the sheep hardly stirring at his quiet walk as he stopped to help at difficult births. He was in time to help a sick ewe to her feet, get her moving before the cold killed her. If a ewe had died, he picked up the lamb before the fox or crows could. At the house, dogs and children get up, watching for him as the frost melts. Has he got a lamb? Has he? No, can't see one.

> "Yes, yes, he has! He's got one!"
>
> And here it came, folded on his arm, its head nodding to his step, coat still wrinkled, yellow and wet from birth.
>
> Now my mother practiced her medical arts. She puffs penicillin powder onto toxic crow pecks, holds the lamb to her while the milk heats up, adds a teaspoon of brandy to the bottle if its tongue is cold.
>
> "This one's not too good. I don't think he's got much chance."
>
> My younger brother and I stood below, looking up at the bump in its throat.
>
> "There! 'E swallowed! 'E did!"
>
> "Look! 'E's drinking!"

Children ousted from their squat against the warm oven door, the lamb was put there on an old jumper, too weak to move or bleat. The audience made up names while they watched and waited. At a point the lamb would push itself up, wobbling on soft, unused hooves onto the old linoleum, open its mouth and bleat in a small voice, "Maa-aa-aa?"

Is there a greater miracle? This one would be all right. Soon there were four or five "poddy lambs" forming a mutually loving company with the children, who learned how to hold the bottle up so the lambs didn't suck air, so the big ones didn't pull the teat off, watching little tails wriggle with pleasure as the warm milk went down and their little bellies rounded. We called them to follow us. They called us if they got left behind, shut out, or it was feed time. The orphans lived with us for months each year, in the kitchen first, then wherever we went outside, until they needed grazing areas, not milk bottles, and joined their cohort of "weaners" in the paddocks.

Young children are usually seen as unable to speak for themselves. In this respect, is their agency more like that of animals? Are lambs equally unable to testify? As Huff comments, "neither non-human animals nor machines possess a human voice or experience their worlds via the technology of natural human language. Luciano Floridi highlights this when he asserts that 'inforgs,' or informational organisms, are agents in both ethical contexts and the construction of personal identities and self-narratives. He removes agency from human exclusivity and locates it instead in the broader category of organism thus granting the latter agential status ethically and as the coaxers of life narrative."[12]

I remember. I can still feel how these idyllic pastoral relationships were intersected by a second flood of experience. Every week or so, a sheep was taken from one little group in the horse paddock to the cow bails, a small shed beyond the machinery sheds. Children weren't allowed to go to the bails, until Les—a small man and an amateur boxer—began his dance around a body hanging upside down on the gambols from the rafter. Les came out from town to do the killing of the sheep because our father hated to. Thanks to him, I learned how to skin a sheep all right.

> "Make a fist with your hand and push here, like this," he instructed, pushing his own fist in between the skin and still-hot sheep's body. The gutter by his feet was lined with congealing blood.
>
> "See? Skin comes away easy."

It did. It was like an open overcoat from the neck to the groin. Its outside was the sweet-smelling gray wool, the inside was warm, embroidered with wet red threads as he threw it onto the top rail to join the others drying there until the Skin Man came to buy them.

"See that line there?" asked Les, addressing the body again with the butcher's knife. His dogs roamed back and forth. "Down the line, open the cupboard, out comes all the treasure." It did too, the paunch looking gray because inside was all green grass, the long, darker intestines mottled with white fat. Heart, liver, and the kidneys in their cocoons of fat went into the bucket.

> "There's the fat y'want to keep for rubbing into y'boots," said Les.
>
> "We put polish on our boots," I replied prissily.
>
> "Ah well." Les picked up the tomahawk, chopped the sheep's head in two and added brains to the bucket.

The sheep was now a barely warm caryatid, with shortened, outstretched arms. He fitted the linen meat bag over it and hoisted it on his shoulder to carry up to the meat house.

"The meat has to set before y'can eat it. Takes a week or so." In time, my father would chop it up and take it to the kitchen fridge in its new reality: cutlets, shoulder roast, leg.

I thought about all the live sheep. I thought about this headless and footless thing in the meat bag. There was something I didn't understand.

> "Can I watch from the beginning next time?"
>
> "No. Y'father wouldn't want y'doing that."

Curiosity armed with ignorance being stronger than obedience, the next time Les came I hid behind the wood heap nearby. Soon, though, I could see nothing, my ears filled with a terrible noise as Les killed that sheep. I would do anything to unhear, to not know, for the rest of my life, and I ran, never far enough away from the rasping for air, desperation for life.

I could never tell anyone what the sheep and Les showed me: how the living sheep became a cold body of meat. Knife, throat, and that terrible noise; once I heard it, I was part Les and part sheep in a young child's lack of screening from experience. It was a gendered experience of its time; unlike my brothers, I never had to learn to kill for meat, castrate, or cut tails. Is that why I obsess over agency, language, and relationship? Without relationship, can there be agency?

Bruno Latour, theorist of material and practical engagement with the earth, argues that "existence" and "signification" have the same meaning. Any being possesses agency, in his reasoning, and agency permeates language: "As long as they are acting, agents signify. This is why their signification can be followed, pursued, captured, translated, formulated in language. Which does not mean that 'every thing in the world is merely a matter of discourse' but, rather, that every possibility of discourse is due to the presence of agents in quest of their existence."[13] And in the case of sheep? Huff proposes that a posthumanist imaginary must "consider the surround in which beings interact."[14] That "surround" is crucial to decentering the human and bringing the other-than-human to equal status in literature.

***

The brief account of sheep in my childhood is blunt and old-fashioned in its homocentricity. I began attempts to decenter the human in my writing and, by chance, was introduced to Schlunke's *Stories of the Riverina.* A third-generation farmer in the German Lutheran community of Temora, his tombstone reads, "Eric Otto Schlunke. Man of Letters, Man of the Land." He wrote in the mid-twentieth century, after the Wiradjuri people's dispossession of the grassy woodlands of southwestern New South Wales, a temporarily stable agricultural environment. As the introduction to the volume argues, his writing is deeply autobiographical.[15] Many of his stories have multiple creatures as active agents in the narrative and the shaping of the text.

In the opening story, "Old Heinrich and the Lambing Ewe," Otto and Heinrich are the only human characters, but crows, horses, dogs, grasses, and the soil itself are characters with motives and personalities that determine events. The actions of the lambing ewe of the title are minutely related and are the ultimate organizing force of the story. The lambing ewe remains the oddly passive focus, a voiceless agent, because the story reflects the enthusiastic naivety of the young farmer. It is not only entities like the birds, dogs, and sheep that are agents in "Old Heinrich," but also cultural understandings in "the multiplicity of existents on the one hand, and … the multiplicity of ways they have of existing on the other."[16] Those "existents" are unpredictable to the human and influence the events, tone, and structure of the narrative: "The crows, which had kept so cautiously out of range yesterday when he had a gun, stayed insolently beside the sheep, walking round it with their long, sharp, black beaks stabbing the air with each step. One even settled on the sheep's head to make a last attempt to maim it by pecking at its eye before Otto, shouting louder and swearing, arrived among them, then it flew away after the rest and settled in a near-by pine-tree to watch, still insolent, as if it were confident it were only a matter of time and it would have the ewe—first its eyes and tongue, then a hole in the chest to get at heart and lungs, helping other crows to strip off the wool and skin bit by bit, quarrelling and fighting, and feasting so heavily each day that it would hardly be able to fly home."[17]

Schlunke's tone is one of affection, tolerance, and a degree of awe at the multiplicity. Some of his (human) characters are strikingly alert to the world as composed of human and non-human involvements and see it as vulnerable. To deviate from sheep for a moment, the story "Hay" is an example of an inclusiveness that determines Schlunke's very narrative technique. The chief protagonist "is" both emu and farmer in the narrative, as he zigzags emu-style in interactions with his farmer neighbors to attain his human goal: getting them to cut his hay. At the same time, the shifting and conflicting priorities of humans and animals of all kinds are shown. The opening reads: "Because the crop was tall and dense, [the Kleinert boys] could see little more than his head, and at first suspected that it was an emu. But emus are prone to stops and starts, to meaningless wanderings hither and thither, and circuitous pursuits of each other by means of which they knock down about fifty times as much wheat as they eat; and the object in the crop was obviously walking straight along a drill wheel-track, which indicated very positively … that it was the owner, taking stock of his crop and not breaking down one more stalk than was necessary."[18]

And it is Schlunke, using what Huff dislikes as "ventriloquized representationality," who wrote "A Sheep's Eye View," in which a sheep narrates its own life story.[19] The nameless sheep is confident, with a sense of its own sophistication that underlines its innocence and naivety. "To the ordinary sheep, all humans are alike. But to an experienced traveller like me, who's been owned and handled by so many different men, they vary nearly as much as sheep," she begins.[20] A good life for sheep does not include being "owned and handled by so many different men." The story immediately alerts the reader to human wrongdoing. Speaking in the present tense, the sheep tells, uncritically, of being auctioned in sale yards, loaded on and off trucks into barren holding yards, and hungering in alien paddocks as dealers, farmers, and auctioneers play the system that is "the market," exploiting each other and the sheep. Although desires, actions, and responses to events are presented as the sheep's, as in all Schlunke's stories, loss of well-being for one means loss for all because animal and human lives are intimately interconnected. At the same time, there are minutely observed details of the country and its ecosystems as the sheep's forced travels continue. The sheep describes itself committing acts that show it becoming a picaresque hero, alienated from other sheep and people alike. It continues to bear witness until its voice and the story stop mid-sentence at the moment when, with the flock, it realizes and experiences slaughter.

"A Sheep's Eye View" does not bring the sheep to language in Huff's sense. For Huff, such a work is "animalography in which human author ventriloquizes the animal's voice allegedly to tell his story."[21] As indeed it is. But again, the structure mirrors a bigger literary dimension: expected order is overturned in giving the sheep language, which in turn mirrors the overturning of the good order of care and respect for the animals shown by the people involved. Can "A Sheep's Eye View" be said to give agency to the sheep? It is a protest story; its metaplot condemns both cruelty and desire for profit beyond need. In her essay devoted to a poem by Judith Wright, Kate Rigby argues that through such literature we may "learn to live in a more just and compassionate pattern of relationship with the more-than-human lives in whose midst, and through whose grace, we ourselves exist."[22] Despite its humorous tone, the details are a realist's account: "Hardly time to get a decent drink, and belly never more than half-full before we started on another trip that might take a day or two or three. On our feet all the time in a lurching, swaying motor-lorry, sometimes with a mug driver who might stop or start suddenly and pile us all up against the front or back of the crate. But to lie down was too risky, particularly if the truck was crowded, as it was apt to be, since those dealers didn't believe in paying a carrier to cart empty space around the country."[23]

Schlunke's stories are "of their time"—pastoral industries and animal husbandry have changed, as has the literature and its critique—but a unique intimacy and understanding inform his focus on the interrelationship of living beings. As I read, I perhaps experience a fuller presence of the sheep itself, either lamb, ewe, or adventurous wether in its intertwined relations with humanity. For my own writing, the stories are rich lessons in locating relations with immediate beings in the perpetual flux of energies and existences. Such literature brings the reader closer to realizing Huff's hope that "interactivity and interdependency presage a new kind of narrative and a new terminology to critique that new narrative."[24]

The Australian National University

Notes

1. Berger, "Why Look at Animals?" 6.
2. Berger, "Why Look at Animals?" 6.
3. Rigby, "Writing in the Anthropocene," 173.
4. Huff, "After Auto, after Bio," 279. Sheep would not want to "deprivilege sight" for themselves. Their eyes are specially adapted, by a long and narrow pupil, always to see the scattered flock grazing around them.

5. Huff, "After Auto, after Bio," 279.
6. Armstrong, *Sheep*, 7.
7. Berger, "Why Look at Animals?" 11.
8. Huff, "After Auto, after Bio," 280.
9. Davison, *Man-shy*.
10. Davison, *Dusty*.
11. Meares, "Whippet," 351.
12. Huff, "After Auto, after Bio," 280.
13. Latour, *Facing Gaia*, 70.
14. Huff, "After Auto, after Bio," 279.
15. Schlunke, *Stories of the Riverina*, vii.
16. Latour, *Facing Gaia*, 35.
17. Schlunke, *Stories of the Riverina*, 3.
18. Schlunke, *Stories of the Riverina*, 22.
19. Huff "After Auto, after Bio," 280; Schlunke, *Stories of the Riverina*, 58–64.
20. Schlunke, *Stories of the Riverina*, 58.
21. Huff, "After Auto, after Bio," 279.
22. Rigby, "Writing in the Anthropocene," 173.
23. Schlunke, *Stories of the Riverina*, 59.
24. Huff, "After Auto, after Bio," 279.

Disclosure Statement

No potential conflict of interest was reported by the author(s).

Works Cited

Armstrong, Philip. *Sheep*. Clerkenwell: Reaktion, 2017.

Berger, John. "Why Look at Animals?" In *About Looking*, 3–30. New York, NY: Vintage, 1980.

Davison, Frank Dalby. *Man-shy*. North Ryde NSW: Angus & Robertson, 1931.

Davison, Frank Dalby. *Dusty*. North Ryde NSW: Angus & Robertson, 1946.

Deleuze, Gilles, and Félix Guattari. *A Thousand Plateaus: Capitalism and Schizophrenia*. Translated and edited by Brian Massumi. London: Athlone, 1988.

Huff, Cynthia. "After Auto, after Bio." *a/b: Auto/Biography Studies* 32, no. 2 (2017): 279–282.

Latour, Bruno. *Facing Gaia: Eight Lectures on the New Climatic Regime*. Cambridge: Cambridge UP, 2017.

Meares, Gillian. "The Whippet and the Willow." In *Collected Stories*, 351–371. St. Lucia, QLD: U of Queensland P, 1997.

Rigby, Kate. "Writing in the Anthropocene: Idle Chatter or Ecoprophetic Witness?" *Australian Humanities Review* 47 (2009): 173–187.

Schlunke, E. O. *Stories of the Riverina*. North Ryde, NSW: Angus and Robertson, 1984.

A Triumphal Entry, a Stifled Cry, a Hushed Retreat

By Rick De Vos

Platypus, Male, Five Years Old, Killed September 3, 1921, The Millstream, Southwest of Ravenshoe, Queensland

I have a secret I'm dying to share. But I dare not now. I know she's here. With me. The darkness is yielding, and I feel her. Near. I search the space around me. Last night an eel, a silent green monster three times my size, swam slowly through the center of the stream, trying to sense my breathing and my movement. My burrow is now no longer safe. Where eels hunt in the stream, snakes follow along the banks, listening for retreating splashes. Before long I will be given up. It's time to move. I look down at the stream, but I can't hear or see her. Just stillness, silence, and the cool, dark water waiting for the light and the birds.

I slide down the muddy bank, watchful and furtive, then close my eyes, ears, and nose as I enter the stream. Last evening, I dug yabbies out of the stream bed not far from here, near the little waterfall. Today, I sense the tiny electric signals from shrimp quietly crossing the stream. They can't evade me. One, two, three, and then a fleeing fourth shrimp fill my mouth as I lunge sideways. My cheek pouches swell. I warily approach the edge of the stream, my eyes and ears open. I eat quickly, then pause to clean the mud off my front feet and stomach. Still hungry. I scan the water in front of me as the light slowly starts to spread over it. Neither shape nor sound. I find the silence strange. Yesterday, the songs of butcher-birds and shrike thrushes rang through the trees above, but not today.

I enter the stream slowly and head to the waterfall in search of yabbies. And maybe her. I pause to scan the branches overhead, against the distant mist. I spy caddis flies hovering above the leaves. I know I'll find larvae below them at the edge of the stream. I swim forward to the bank, pouncing and digging swiftly through water, leaves, and sand, my bill scooping up dozens of tiny grubs. Two more dives bring me more larvae and a few large worms. I clean myself off and again take to the stream, following the spray and pulse of the waterfall. In the water, I sense a stirring of life: small fish gleaning the rocks and pebbles, frogs making watchful crossings in leafy shadows, the sudden, sporadic dives of kingfishers. I move on, feeling the growing force of the stream's current as I near the waterfall.

This is my dilemma. I'm scared that if I speak my secret, it will disappear. Best not to get your hopes up. Just forget about it. I stop and move to the bank as I feel the surge of rolling, cascading water. My eyes and ears are filled with this confusion of light and noise, but as I adjust, I sense shrimp close by. Seizing my chance, I plunge in the direction of the movement and am rewarded with several plump morsels, which I squeeze into my cheek pouches. Emerging from the foam, I scan the banks for movement. The little waterfall was where my mother brought my sister and me to practice hunting. She taught us how to avoid water rats and to sense the presence of snakes. Once, while following our mother out of the stream, we saw her stop dead, almost rearing back. She hurriedly jostled us back into the water, but not before I caught a glimpse of silver netting glinting in the sun, and a small, dark mass of what I think was fur and blood. We swam well up the stream before crossing to the other bank, and never returned to that spot while we were together.

The banks along this side of the stream are deserted. I can see several openings along the surface. I remember at least one or two older males lived near here. I learnt to wait until they had retreated to sleep before digging in the stream. Now, all I see are homes gone to ruin and a space marked by emptiness. I want to feel hope. I want to feel safe. Happy. But I am gripped by the eeriness of this place, and by a sense of dread. I try to shake this off, and select a sheltered opening. I start digging, making the burrow just large enough for me to rest in, flattening my feet to smooth the walls around me. Peering out, I imagine I can see one or more large, pale shapes moving beyond the other side of the stream, but in a second they are gone. I venture out of my new dwelling slowly, and slide into the stream. Digging has made me hungry, and I dive for the stream bed, quickly scooping up a few pea clams. I pull them apart and devour the fleshy pods within. I swim toward the waterfall, testing the stream for shrimp and yabbies, seeking to satisfy my hunger so that I may return to my new home and sleep.

And then I see her. Through the mist and spray she stands on the bank, her eyes sweeping the stream. She dives, and I swim cautiously in her direction, pausing to see if she is still near. She has stopped, too, regarding me deliberately before moving toward the stream's edge. There is fire in her eyes, a burning spark that I feel inside me. She knows my secret. I am happy. This time, I approach swiftly and purposefully. She waits for me. Our eyes meet and, for an instance, we are frozen in time, facing each other. Her presence fills and overwhelms me. Our bodies brush against each other, and I follow her into the swirling stream, taking hold of her tail with my bill. She twists, then turns to bite mine. We roll together, the two of us tumbling against the current of the stream. Above me, I am faintly aware of a tall, pale shape against the bushes by the stream's edge, but I am intent on staying with her, following the course we have begun. I move my front feet onto her back and slide over her. I see nothing but her; I feel nothing but her. She is my world.

CHOOOM!!!

Her body convulses, then hangs limp. I taste and smell blood and, as I lift my head up, smoke. The tall, pale shape moves toward us, entering the stream, with eyes that have no fire, eyes that are cold and dead. Confused and in shock, I back away, and then, as I see this grim destroyer attempt to take hold of her body, my anger surges and I hurtle toward it, seeking to drive it away.

CHOOOM!!!

I hit the water, no longer in control of my body. I cannot breathe. I feel my body burning, even as the cold stream pulls me under, and then, nothing at all. Just that she is here with me.

> I witnessed a pair copulating. The pair were in an eddy of a swift current and the male slid over the female, from the rear, and assumed a position in which he could very easily stick his spurs into the sockets of the female. I was standing on the bank, only about five feet from them, but owing to the swirling of the water and the way in which they bobbed about, I was unable to see whether he actually used his spurs or not. I subsequently shot the female, and while I was getting her out of the river, the male came to the surface and swam up to within three feet of me. Swelling out his body and floating high, until he was almost walking on the water, he threw his head from side to side, hissing and vigorously splashing the water with his feet and tail. This seemed unusual behavior for they are generally extremely shy. The cheek pouches examined contained aquatic insects, worms, prawns, and small shellfish. Although they have a bad reputation among fisherman as fish eaters, I never found the slightest sign of fish or fish spawn in their cheek pouches.[1]

The Vertebrate Zoology Collection of the US National Museum of Natural History, administered as part of the Smithsonian Institution, includes in its holdings a number of complete and partial specimens of platypus, Ornithorhynchus anatinus. They include the specimens depicted in figures 1 and 2, *collected by the American specimen-collector Charles McCauley Hoy (1897–1923) on September 3, 1921, five miles southwest of Ravenshoe, Queensland, which at the time was a timber town that had only been connected five years previously to the commercial center of Cairns by rail. The area surrounding Ravenshoe in the Atherton*

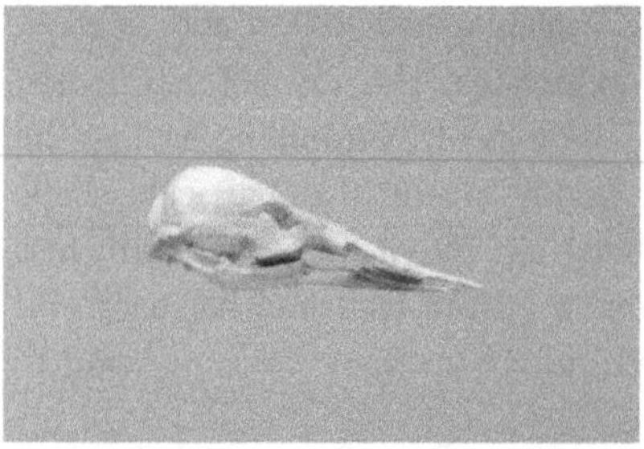

Figure 1. *Skull of female platypus,* Ornithorhynchus anatinus. *Collector: C. M. Hoy, Atherton Tableland, five miles southwest of Ravenshoe, September 3, 1921. Catalogue number: USNM 238452. Department of Vertebrate Zoology, Smithsonian Institution. Photograph by Donald E. Hurlbert.*

Figure 2. *Skull of male platypus,* Ornithorhynchus anatinus. *Collector: C. M. Hoy, Atherton Tableland, five miles southwest of Ravenshoe, September 3, 1921. Catalogue number: USNM 238451. Department of Vertebrate Zoology, Smithsonian Institution. Photograph by Donald E. Hurlbert.*

Tablelands was characterized by rainforest, waterfalls, crater lakes, and swamps. European settlers hunted platypus for fur in the area in the latter half of the nineteenth century, leading to an observed decline in platypus numbers, at least until the Queensland government introduced legislation in 1906 banning the hunting of platypus.

The physician, explorer, and field naturalist William L. Abbott, a significant sponsor of the museum, arranged for Hoy, who had been serving in the US Navy, to be discharged in order to travel to Australia in 1919, with the express purpose of collecting vertebrate specimens, in particular those in danger of extinction.[2] *It was Abbott's intention that Hoy, who had received training in fieldwork and specimen collection at the museum, should make the initial foray to Australia in anticipation of the arrival of senior field naturalists and curators William Gregory and Harry Raven, who were resting after expeditions to Celebes and Borneo. The trip was prompted by the museum's fears that a number of Australian mammals, including platypus, were close to extinction, and that their own holdings of Australian mammals were sparse and risked never being completed. Hoy was dispatched to Sydney in May 1919, and spent nearly three years traveling around the country, moving between metropolitan areas by coastal ships and using local available transport in rural areas. He consulted local museums in order to gain specialist knowledge and to contact local mammal experts to guide him in his search for suitable areas. During his time in Australia, Hoy collected over one thousand mammal specimens from approximately one hundred species, taken from fourteen identified sites. Hoy used field notebooks to record the time, place, and zoological details of each specimen, cataloguing them chronologically as per the museum's instructions. Hoy's initial assumption, reinforced by his first year in Australia, was that numbers of most native mammal species were rapidly declining due to a convergence of factors, including invasive species and land clearing.*

Brian Massumi contends that the notion of threat, as evidenced by events in the past, produces a preemptive logic operating recursively between the present and the future, and on an affective register. Threat elicits fear even when no material evidence occurs in the present; indeed, Massumi defines fear as "the felt reality of the non-existent."[3] *For the US National Museum of Natural History, the fear of species extinction in Australia appears to be subjugated to the fear of a museum collection being left with a gap that cannot be resolved. The two fears are conflated as a single threat, a foreshadowing of a time when the species in question have all but disappeared and it will no longer be possible to collect the specimens required to complete the museum's collection. However, as Massumi argues, the affective power of threat enables it to write over the reality of a specific site or context with its own conditional determinism.*[4] *For the museum and its agent Hoy, collecting thus becomes a compelling mission, one without defined limits, and simultaneously constitutes an activity that runs the risk of making the threat of extinction manifest.*

The fear that compels this preemptive action is not felt for Australian mammals, but for the possibility of a loss that might be felt by the museum, by its staff, stakeholders, and clients. What is at stake is not the end of interconnected ways of life, of knowledges and experiences of survival and endurance—in short, what is understood in a narrower sense as "species"—but rather of specimens, those products of killing and violent extraction that afford their exhibitors varying degrees of social status, economic wealth, and scientific gravitas. And yet, the fact that they are displayed and exposed provides an opportunity for other stories to be told, for other perspectives and lives to be revealed, and for new narratives to retrieve and explore the absent experiences and ways of being.

* * *

> Thus the chance shot of mine altered my career, and, at the same time, supplied science with the knowledge of a new and most interesting animal.
> — Charles M. Hoy, "The 'White Flag' Dolphin of the Tung Ting Lake"

Baiji, Male, Killed February 18, 1914, Dongting Lake, Hunan Province

As always, we dived as we sensed the boat in the distance, its echoes beating dull and hollow, devoid of life, ushering all away. The fluttering silence of fish disappearing toward refuge and the high-pitched clicks of *ju*, sounding warnings and threats, were all that met us in the pale midmorning shadows. We moved together, heading south toward the swirling waters where the lake met the Xiang Jiāng. And then I paused and swam up, compelled to breathe, breaking the surface momentarily. As I made to dive again, my body shook and stopped. I fought for breath, gasping and choking as I sank helplessly, the cool water burning me, the darkness enfolding me in the midst of the morning light.

* * *

The lake was not merely a place: the lake was life. From the south, the currents of the Xiang Jiāng brought shimmering ribbons of silver fish seeking shelter along the lake's edges. To the north, we would follow the currents as they emptied into the middle reaches of the swift-flowing Cháng Jiāng, setting out on hunting forays, joyfully pitting our bodies against the river's steady flow and menacing undercurrents. The depths of the lake's center, however, were from where we set out, to where we returned, and where we dwelt at night. It was where we occasionally encountered others, other families and visitors, travelers from other waterways, and where we formed new groups and made new families. The lake was lively and dynamic, flooding in the rainy season as it received the runoff from the Cháng Jiāng and its channels, and receding in the dry season as the lake's shores were exposed.

We learnt to wait for the fish each year, halfway through the rainy season, circling the rocks and sandbars near the entrances to the long river, watching for the waves of silver carp and grass carp as they rushed into the lake and were sucked up in the swirling eddies. We grew fat on these fish, seizing them from below as they paused, lost in the void created by the current. At the end of the season, as the flood plain emerged and the carp grew scarcer, it was time for us to journey into the river, swimming for many miles in search of new hunting grounds. I learnt from my family how to negotiate the currents and locate the points of confluence that would provide the best chance to find fish. We avoided places where we smelt smoke in the air, where fish were already dead, and stayed well away from all boats and large objects. Through the course of the seasons, we learnt to fine-tune our sense of the space around us, the river banks, channels, and meanders, and, more importantly, our sense of each other and of how to anticipate each other's movements. We communicated quickly and more urgently than we did in the lake, our less familiar surroundings uniting us in our endeavors.

Eventually, however, we would return to the waters of the lake, to its relative tranquility and safety. We shared the lake with *ju*, our smaller, noisier neighbors, who swam close to the surface and lived off smaller prey. We often followed their movements from below, their dark bodies scurrying above us, their high-pitched chattering alerting us to approaching fish and to possible danger. Sometimes we swam together, patrolling the lake's vast borders, learning over time the vagaries of the channels, sandbars, and currents, becoming aware of changes in the movement of the water, and of the movement of fish and other lake inhabitants. Our patrols also served to keep us aware of movement on the shores of the lake, and particularly that of humans, forever increasing in number and activity, their smell and their refuse tainting the edges of the lake, their presence reinforcing our preference for the deeper water in the center of the lake. We avoided them for other reasons, too. We shared stories and memories of our companions disappearing, and suspected fishermen and their families, working surreptitiously before the first light.

* * *

Two articles by Hoy appeared in 1923. The first, published in the China Journal of Science and Arts, detailed his encounter with a baiji, or Yangtze River dolphin, in 1914, while Hoy was living with his family in Hunan Province in China, where his father worked as the principal of a Christian seminary college at Yueyang.[5] *Hoy had been on a duck-hunting expedition on Dongting Lake when he spotted a group of ju, or Yangtze finless porpoises, and baiji swimming away from the boat. He shot and killed a baiji, which was brought to shore by local villagers. After the baiji had been cooked and eaten, Hoy retrieved and cleaned the skull, along with a couple of cervical vertebrae. The following year, he returned to the US and sold the skull* (figure 3) *and vertebrae to the US National Museum of Natural History. The baiji was subsequently described as a new genus and species of river dolphin, and Hoy was offered a place on the US Biological Survey, where he received formal training in specimen collecting.*

Hoy's second article was published in the Journal of Mammalogy, and provided an overview of his time in Australia.[6] *Despite his earlier warnings about the plight of Australian mammals, Hoy concluded that the threat of extinction was in most cases overstated, largely as a result of sampling from a limited range of locations. This opinion echoed the more specific sentiments he expressed in his field notes in regard to platypus. He expressed the hope that if the Australian government continued to enforce closed hunting seasons and establish and maintain sanctuaries, most species could be saved. This opinion was absent from the US National Museum of Natural History's 1922 annual report, which detailed Hoy's list of specimens collected. By the time Hoy's articles had been published, he had died. Having returned to China to conduct work for the US National Museum, he developed an appendicitis, aggravated by a blood-fluke infection he most likely contracted as a teenager near Dongting Lake. He died on September 7, 1923, aged 26.*[7]

In examining the representation of encounters between humans and nonhumans, Kathryn Yusoff identifies the ways in which such representations are shaped by images, archives, and institutions such as zoos and museums. Yusoff argues that these representations, the result of aesthetic and ethical practices, are directly related to loss and extinction, and subject creatures to the sovereignty of biological science.[8] *Part of this practice involves separating the subject's representation from the violence involved in producing this subjectivity. The degree to which biotic subjects are subjectivized is influenced by what is seen and what remains unseen. Cryptic animals like baiji are more given to nonrepresentational acts, bringing resistance to this practice of subjection|representation, and are less likely to represent an ethical demand and to be afforded protection. Yusoff asks us to consider the conditions under which nonhuman animals become subjects. In the case of baiji, their enunciation as species by way of a skull and vertebrae obtained through an act of killing, and through consistent representations as rare and elusive, reinforces their status as vulnerable. Extinction is inscribed in their appearance both as specimen and as species. Nearly every photograph of a baiji depicts Qi Qi, the male baiji kept for twenty-two years at the Wuhan Hydrobiological Institute, who was consistently described as the last baiji in the world. The captive Qi Qi, to quote Yusoff, can be seen as "caught between two worlds, as a life held in a space in which it cannot be held."*[9] *The challenge remains, in writing baiji lives, to resist the predetermination of extinction and imagine lives that transgress this state, lives before and lives beyond this subjectivizing temporality.*

* * *

Now the lake is no longer recognizable; indeed, it is no longer the extensive waterway we lived and swam in, and that connected the Xiang Jiāng and the Cháng Jiāng, beckoning us to venture beyond and to return with the passing of the seasons. Now, three distinct bodies of water lie in its stead, segregated by wide expanses of mud and silt. The lake edges have receded, the land reclaimed for human habitation and farming. The clean, clear water of the lake we knew has been transformed into a turbid mix of browning suds, human and other animal waste, fertilizers, and machine oil.

The remnants of my family left the lake some years after I had gone. Humans with whom we had shared the lake and abided for generations now actively targeted and killed us, stunning us with deadly electrical charges or trapping us on long lines of countless metal hooks. Those who survived had no option but to flee the lake, no longer a refuge but a stinking, noisy wasteland where fish could no longer be found and where our family and neighbors started to sicken as they swam along the poisonous banks. The hardy *ju* remained for a while after we had left, perhaps imagining they might profit from our absence, but in time they, too, were compelled to quit the putrid, stifling waters that had once sustained them.

The Cháng Jiāng, however, was not the sanctuary they had hoped for. Where once we negotiated the river and its channels, locating confluences and eddies, my family was met with a waterway congested with boats and human activity. They dived to seek shelter and make sense of this new world, but neither was possible. The boats were many more in number, noisier, and faster. The noise from the huge machinery lining the river was equally deafening. Locating echoes became impossible, as was attempting to communicate directly with each other. Family groups became separated and disoriented, and individuals were battered and killed in collisions with the boats that ceaselessly plied the river. The opportunity to teach each other to negotiate the river, to hunt, to work together, and to learn from each other, was lost.

Worse was to come. As individuals sought to find food and sanctuary further along the river, they were met with walls and barrages. The river's currents were arrested, diverted, and blocked by dam walls and turbines. The cumulative effect of these disturbed and unfamiliar flows, the wake from countless water vessels, the noise from industrial machinery and boats, and the polymorphous pollution that pervaded the river was both blinding and deafening. Our way of life had been halted, our interactions occluded, our sense of the world obliterated. There was nothing to see, nothing to hear, nowhere to go. My family's experiences, knowledges, and sense of meaning incrementally disappeared with each individual's death.

We are finished; we are no more. I wonder still what purpose it served for us to be annihilated. Simply for sport? Purely for future generations of other beings to stare in comfort and curiosity? Merely for our collected bodies to remain in stasis? One unfortunate individual was captured by humans: he lingered for twenty-two years in a concrete pool. They considered him the last of his kind. But we were already gone. Without each other, without the seasons, without the lake and the river, we are merely bodies on display waiting to die. They thought they could save us, but they did not know us. We had a past and a future, generations of knowledge and experience, lessons learnt, obstacles overcome, grief and happiness felt and shared, but this was not comprehended. They never made the time and never made the space. What was there to save? Any memories of us were corrupted and inflected through guilt and the desire for redemption. Yet our lives and deaths mattered to us and to those with whom we lived and interacted. The world was different because of our presence and is different because of our absence.

* * *

Once more I return to that morning, to that moment of one death that foretold the death of many. A lake full of life, soon to become void. Once again I feel my companions around me, feel the current against my body.

I sense the echoes of the boat and dive below, clinging to a last inkling of hope. And then, as always, I rise for a last breath, surrendering to my fate.

> [A]fter waiting a few minutes, a single animal suddenly made its appearance about 70 yards from the boat. Although this was rather a long shot for so small a target, the dolphin only showing a very small part of its body above the water, I took a chance shot at it, and turned away to lay my rifle down in the bottom of the boat, feeling certain that I had not made my mark. However, one of the natives shouted out that I had hit the animal, and, on looking round, I saw its snout sticking up above the water. I realized at once that what the native had said was true, and immediately gave orders to paddle over to the wounded creature. As we neared the spot, the dolphin gave out a subdued bellow, somewhat after the nature of the noise made by a buffalo calf. We reached the animal just as it was sinking. It was evident that we were going to have a great deal of trouble in securing our prize, since the body was very slippery and there was little or nothing to lay hold of, but one of the boatmen, watching his opportunity, grabbed the creature by its long snout. This was a very risky thing to do, since the jaws of the animal, which are armed with series of sharp teeth, were snapping vigorously. The man had to wait his opportunity and seize them while they were closed; otherwise he would have had his hands badly lacerated. He hung on like grim death, giving the rest of us a chance of slipping a rope round the body of the wounded animal and fastening it just behind the flippers. It was too heavy to lift into the boat, so I gave orders for the crew to paddle for the shore—about a mile away—with the dead monster in tow. On arrival there, we carried it back to the college, making a triumphal entry.[10]

Figure 3. *Skull of male baiji,* Lipotes vexillifer. *Collector: C. M. Hoy, Dongting Lake, Hunan Province, February 18, 1914. Catalogue number: USNM 218293. Department of Vertebrate Zoology, Smithsonian Institution. Photograph by Donald E. Hurlbert.*

Curtin University

Notes

1. Hoy, "Field Notes."
2. Hoy, "Field Notes."
3. Massumi, "Future Birth," 54.
4. Massumi, "Future Birth," 56.
5. Hoy, "'White Flag' Dolphin," 154–155.
6. Hoy, "Present Status."
7. Turvey, *Witness to Extinction*, 10–11.
8. Yusoff, "Aesthetics of Loss," 582–583.
9. Yusoff, "Aesthetics of Loss," 583.
10. Hoy, "'White Flag' Dolphin," 156.

Disclosure Statement

No potential conflict of interest was reported by the author.

ORCID

Rick De Vos http://orcid.org/0000-0002-5738-9518

Works Cited

Hoy, Charles M. "Charles M. Hoy Field Notes, 1919–1923." *Smithsonian Institution Archives.* Accessed November 9, 2019. http://siris-archives.si.edu/

——. "The Present Status of the Australian Mammal Fauna." *Journal of Mammalogy* 4, no. 3 (1923): 164–166.

——. "The 'White Flag' Dolphin of the Tung Ting Lake." *China Journal of Science and Arts* 1, no. 2 (1923): 154–157.

Massumi, Brian. "The Future Birth of the Affective Fact: The Political Ontology of Threat." In *The Affect Theory Reader*, edited by Melissa Gregg and Gregory J. Seigworth, 52–70. Durham, NC: Duke UP, 2010.

Turvey, Samuel. *Witness to Extinction: How We Failed to Save the Yangtze River Dolphin*. Oxford: Oxford UP, 2012.

Yusoff, Kathryn. "Aesthetics of Loss: Biodiversity, Banal Violence and Biotic Subjects." *Transactions of the Institute of British Geographers* 37, no. 4 (2012): 578–592.

Her Biography: Deborah Bird Rose

By Stephen Muecke

Her Geography

If you are driving west from Katherine on Highway 1, headed into the Kimberley in the far northwest of Australia, take time to pause and look south down the Buchanan Highway; that's the turnoff before you get to Timber Creek. Debbie asked me to just nod my head, once. "Say hi to the people at Yarralin for me; I haven't been able to get up there lately."

Yarralin, on Victoria River Downs, is about an hour's drive south of the turnoff, and Victoria River Downs is one of the biggest cattle stations in the world. The mob there wouldn't have seen me nod. Or you, if you remember to do it, if you ever get to travel the country she once walked, that she came to love in so many ways.

She was born in Orono, Maine, on October 29, 1946. Much of her childhood was spent in Seattle, Washington, in the Pacific Northwest, the region she considered her homeland in the US and also a region with very strong Native American traditions. It was her brother, Will Rose, who told me that "the Pacific Northwest [was] the region she considered her homeland in America."[1] By adding "in America" he knows that she also made Australia her home from 1980, and Yarralin, I'm sure, was another homeland, though she lived and worked in other Aboriginal countries, mostly in the Northern Territory. City life, where her university work was carried out, was mainly in Canberra and Sydney.

According to her brother Will, the early years in her family of five children were marked by frequent moves urged by her restless father: from Orono to Seattle, to Wyoming, Oregon, and Massachusetts, to France, and back to rural Massachusetts. Debbie's year in France with her family, from 1962–1963, seemed to open her eyes to other cultures and languages, setting her up for later work, no doubt. As a student at the international Collège Cevenol in Le Chambon-sur-Lignon, France, where her family spent that sabbatical year and to which Deb returned briefly as a teenager, she was exposed, as Will remarked, "to a post-war ethic of compassion for others, pacifism, multiculturalism, and ecumenism."[2]

Her Anthropology

Strange as it may sound, Debbie was brought up with the prospect of everlasting life. But that is Christianity for you. Her father was the Reverend David D. Rose, and both her grandfathers were liberal Protestant ministers in the Congregational and Unitarian denominations. In her heritage was a deep respect for spiritual life—a respect that did not go astray in Aboriginal Australia. Here, too, she learned that life was everlasting, but there was another valuable lesson she learned from her Aboriginal teachers—salvation was not for individual lives as such, but life forces were distributed among all sorts of living things in the environment. Life didn't so much go *up* (to Heaven) or *down* (to one's descendants). Rather, it was forever suspended in horizontal networks that kept everything alive. She would have heard the phrase "generation after generation" many times, often expressed with a "tumbling-hands" gesture that indicates cyclical life rather than descent.

Deborah Bird Rose became a flying-fox woman after she went to Yarralin. I have to talk a little about totemism and what that concept means, because she revived the concept of totemism as a shared-life concept that distributed rights and responsibilities for ecological management. It came up in the last communication we had via email on September 2, 2018. "Astonishing, isn't it," she said, "that the one big in-your-face fact—that 'totems' are your kin in a world of sentient beings—just never hit the horizon!" When Debbie came to Australia in 1980 to do the doctoral work that became *Dingo Makes Us Human*, she would have been well aware that she was landing on a continent whose cultures had long been characterized as "totemic," initially via what Émile Durkheim derived from his reading of Walter Baldwin Spencer and Francis James Gillen. The word "totem" is not indexed in *Dingo Makes Us Human*, but the term for "matrimoieties," she writes, among the people at Yarralin, is *ngurlu*: "*Ngurlu* is a relationship between people and one or more plant or animal species. People of the same *ngurlu* share the same flesh with each other and with their *ngurlu* species."[3] This is a fairly standard definition of a totemic relationship—beings of different appearance but sharing the same life force. And we learnt that Deborah's *ngurlu* is flying fox in a matrilineal line: "My skin is Namija. I was also assigned flying fox *ngurlu* on the basis of my close relationship with my 'sister,' Mirmir. She is flying fox because her mother and her mother's mother are flying fox."[4] But "the particular species with which one's *ngurlu* is identified is called one's *warpiri*," and there are especially intense relationships associated with them.[5] Wisely, Deborah resisted conflating *ngurlu* and *warpiri* under the heading of "totem" because these concepts that inalienably link humans and nonhumans are plural and complicated. The flying fox thus shared her life and extended her life into the environment in a logical way—a logic of care, rights, and responsibilities. She finished her flying-fox book, *Shimmer*, just before she died. It will come out in 2020.

She was the first anthropologist I knew who referred to the people she worked with as her "teachers." One could say that the older identity of "informant" was one of those many roadblocks to decolonization. It is a subaltern category a long way from "teacher" or "colleague," which more enlightened anthropologists started to use long after her—subaltern because the anonymous "informant" is a name that functioned like a valve in settler-colonial knowledge acquisition. It helped transform Indigenous knowledge into something acceptable on the way out, and allowed nothing to be transformed on the way in. Thousands of these valves kept the local knowledges local, while feeding the spread of sciences and social sciences around the country. The modern sciences and social sciences were thus, with the help of knowledge valves like this, able to continue their universalizing mission. But as soon as Debbie started using the word "teacher," you had to understand that people like Hobbles Danayarri ("one of the most philosophically gifted of Yarralin and Lingara historians," she writes)[6] were actively transforming what people like Debbie thought she knew, and by extension the discipline of anthropology in Australia.

Her Poetics

Her student Stuart Cooke, writing an appreciation in a forthcoming Festschrift, reminds me that Debbie had said, in dialogue with the great English Australian poet Martin Harrison, that "ecological writing does in the text what life does in ecological connectivity."[7] She was, in that respect, a creative writer. She was a great admirer of Aldo Leopold, Rachel Carson, and many others, and wrote with their kind of care and devotion about things that really mattered, knowing that the means of communication are as important as the knowledge itself. A poetics is the expression of a form that "hangs together" (like bats!), a creation that has been *instaured* and doesn't unravel with time. Judgments about "what matters" are made—they have to be made—but they are less important than making connections as organically as possible, with such craft that the ideas fly out on strong wings on the haphazard breeze, to roost in trees, shitting inadvertently on shitty ideas—Debbie would laugh—before echolocating once again, making sense once again.

Her Storytelling

But stories, articulated with time that passes, are a good device for teaching and learning, especially while walking somewhere. Here is Debbie learning about flying foxes (*warrpa* in Yarralin): "One of my teachers was Daly Pulkara, a man with a good fund of flying fox stories. We were watching the flyout one evening, and he pointed out something interesting. The crowd flew over, and then, after the mob was well on its way, a few turned back. A bit later, stragglers appeared, following the others but not quite catching up. 'The old people always said those blokes forgot their axes,' Daly told me. I felt like a bit of a straggler myself as I tried to catch the drift of the story. Of course, they were shape-shifters, and when they were men they would have carried axes. 'They're always like that,' Daly said, 'one or two are back behind. That's why the old people said they forgot their axes, they had to go back to camp and get them.'"[8]

In her appreciation of Debbie's work on flying foxes, Tessa Laird writes of how the chattering of the foxes wakes up the Rainbow Serpent, urging him to bring rain at the end of the dry season: "With suitably floral language, Rose calls the pollen-loving flying foxes 'co-evolved myrtaceous mutualists.'"[9] But calling them "flying foxes" is even stranger to Australian ears, since foxes are an introduced species, and now something of a rogue species, while many Australians also find flying foxes noisy, smelly, and pestiferous, not appreciating how they are "endangered, and are involved in all four of the major factors causing extinctions: habitat loss, overexploitation, introduced species and extinction cascades."[10] This is why Debbie calls them a "keystone" species, symbiotic with the lives of others: the Myrtaceae family of trees, in totemic relations with clans of humans, and in what we call a "spiritual" relationship with the Rainbow snake, not having a better word.

In March 2014, Debbie wrote, "Rachel Carson's *Silent Spring* continues to haunt us with remembrance of all the Earth is losing. It calls us repeatedly to realise how beautiful are the lives of others, and how precious. With remembrance comes sadness." She finishes that blog about song and songbirds with a sentence that I wish I had written about her, for this piece, as a "singer of new songs," but she was always one step ahead of me: "Amongst the great and varied kindred of Earth life, blessed are the singers of new songs—they bring creativity, along with all this great wild beauty, to the symphony of life."[11]

Flinders University

Notes

1. Personal communication, May 2019.
2. Personal communication, May 2019.
3. Rose, *Dingo Makes Us Human*, 82.
4. Rose, *Dingo Makes Us Human*, 82.

5. Rose, *Dingo Makes Us Human*, 83.
6. Rose, *Hidden Histories*, xxii.
7. Cooke, "Poetics." The dialogue he refers to is Harrison and Rose, "Connecting," 1.
8. Rose, "Flying Fox," 122.
9. Laird, *Bat*, 148.
10. Rose, "Flying Fox," 119.
11. Rose, "Songsters."

Disclosure Statement

No potential conflict of interest was reported by the author.

Works Cited

Cooke, Stuart. "The Poetics of Deborah Bird Rose." In *Kin: Thinking with Deborah Bird Rose*, edited by Thom van Dooren and Matthew Churlew. Forthcoming.

Harrison, Martin, and Deborah Bird Rose. "Connecting: A Dialogue between Deborah Bird Rose and Martin Harrison." *TEXT* 20 (2013): 1–5. http://www.textjournal.com.au/speciss/issue20/Harrison&Rose_Postsc.pdf

Laird, Tessa. *Bat*. London: Reaktion, 2018.

Rose, Deborah Bird. *Dingo Makes Us Human: Life and Land in the Australian Aboriginal Culture*. Cambridge: Cambridge UP, 1992.

Rose, Deborah Bird. "Flying Fox: Kin, Keystone, Kontaminant." *Australian Humanities Review* 50 (2011): 119–136.

——. *Hidden Histories: Black Stories from Victoria River Downs*. Canberra: Aboriginal Studies, 1991.

——. "Songsters." Deborah Bird Rose, Life at the Edge of Extinction. Monthly Archives, March 2014. Accessed November 11, 2019. https://web.archive.org/web/20140828141825/http://deborahbirdrose.com/2014/03/

Artist's Statement

By Anna Laurent

Flora: Exploration to Extinction is an immersive installation that reevaluates our relationship to the natural world. A collaged landscape of flora displays an ecological Utopia of abundance, referencing a nineteenth-century period of botanic discovery and a cultural perception of limitless global resources. The endless lush rainbow is interrupted by silhouettes of extinct and endangered species, a prompt for viewers to recognize the reality of contemporary threats to our global biodiversity. *Flora* is a site-specific public installation in London's South Kensington.

Each of the black silhouettes is accompanied by an information panel including the plant name, native ecology, and ecological pressures. Black silhouettes were chosen as a visual cue for absence in the landscape, and as a reference to Victorian portraiture and nineteenth-century plant exploration, during which species traveled the globe and luxurious images of flora were fetishized. The silhouettes are highlighted by gilt frames, emphasizing the portrait presentation and nostalgia for the nineteenth-century adoration of flora, as well as honoring threatened species and portraying them as family. Dried seed heads and leaves, spray-painted gold, are wound around the frames, plunging from the walls as preserved relics and living plant matter that cultivates an immediacy and intimacy between the installation and its audience.

Anna Laurent is an award-winning flora-focused writer, photographer, and artist. She is especially interested in the visual depiction of plants—as an author and historian, throughout history, and as an artist, in contemporary practice. *Dispersal*, a series of photograph portraits exploring the form and function of seed dispersal, has exhibited internationally at the Royal Botanic Gardens, Kew, London, the Arnold Arboretum of Harvard University, and the University of California Botanical Garden at Berkeley, as well as gardens in Amsterdam, Italy, Gibraltar, Sydney, and New York City. In 2017, *Dispersal* was awarded a silver-gilt medal by the Royal Horticultural Society for a selection of works juried for the Chelsea Flower Show. In 2018, the Royal Botanic Gardens, Kew commissioned a series of seed portraits to be permanently displayed alongside living plants in the Temperate House, the largest Victorian glasshouse in the world.

Her first book, *Botanical Art from the Golden Age of Scientific Discovery* (University of Chicago Press, 2016), examines the art and science of plant biology in nineteenth-century wall charts. A botany primer with contextualized background, the book emulates a Victorian classroom while providing biographies of renowned illustrators and scientists.

Born in Boston, USA, Laurent studied at Harvard University and the Massachusetts College of Art and Design. She holds a Master of Science in History and Philosophy of Science from University College London. Currently, she lives in London and has been awarded a UK visa as an Artist of Exceptional Talent.

www.annalaurent.com

Index

Figures are indicated by italics. Endnotes are indicated by the page number followed by "n" and the endnote number e.g., 240n15 refers to endnote 15 on page 240.

www.ingramcontent.com/pod-product-compliance
Lightning Source LLC
LaVergne TN
LVHW011117300926
849160LV00008B/423